Lecture Notes in Mathematics

A collection of informal reports and seminars
Edited by A. Dold, Heidelberg and B. Eckmann, Zürich

253

Vadim Komkov
Texas Tech University, Lubbock, TX/USA

Optimal Control Theory for the Damping of Vibrations of Simple Elastic Systems

Springer-Verlag
Berlin · Heidelberg · New York 1972

AMS Subject Classifications (1970): Primary: 49 B 25 Secondary: 73 C 99, 73 K 05, 73 K 10

ISBN 3-540-05734-X Springer-Verlag Berlin · Heidelberg · New York
ISBN 0-387-05734-X Springer-Verlag New York · Heidelberg · Berlin

Offsetdruck: Julius Beltz, Hemsbach/Bergstr.

FOREWORD

This monograph intends to fill the existing gap in the applications of optimal control theory to problems of damping (or excitation) of simple elastic systems. Some of the material follows closely the contents of articles concerning the control of hyperbolic systems of D. Russell and of articles of the author concerning the control of beams and plates. Some of the material has never appeared in print before. Some obvious generalizations have been omitted, but some more difficult generalizations, such as the control of a vibrating arbitrary three dimensional elastic body, have not been solved yet.

This monograph is intended to be a self-contained exposition of the basic principles of optimal damping of vibrations of simple elastic systems. The reader is assumed to be familiar with advanced calculus, some elementary concepts of functional analysis and some concepts of partial differential equations. For the sake of convenience the author includes a basic discussion of admissible distributional controls in Appendix 1, and an expository discussion of the classical form of Pontryagin's principle is offered in an appendix.

List of Contents

0. INTRODUCTORY REMARKS

The study of the behavior of vibrating elastic systems goes back to Jacob Bernoulli who established the differential equation for the deflection curves of elastic bars [1], and to Leohard Euler who investigated the vibration of a perfectly elastic membrane (see [38], Chapter 2). The differential equations of vibrating thin plates can be traced to Kirchhoff (see [38], Chapter 8).

The control of vibration (or damping of elastic vibrations) has long been considered an important problem, and various experimental techniques are quoted in engineering papers as far back as early 19th Century. In his translation of the book of Clebsch, Saint Venant gives an early account of a control problem for a vibrating bar. Other authors who considered this problem include Philips, Boussinesq, and Duhamel. It is also Duhamel who developed the basic differential equations of non-homogeneous vibrating plates.

A historical outline of the ideas in the theory of vibrations of beams and plates is contained in Timoshenko's book [38]. Additional information may be found for example in the introductory notes of Love 's treatise [21]. A modern approach to the mathematical theory of control for vibrating elastic systems based on Pontryagin's ideas [28], [29], has been originated in

Russia by Butkovskii, Lerner, and Egorov. ([4], [5], [6], [7], [11], [12]).

In [10] Egorov considered the control problem for the heat equation, and in [11] in cooperation with the other authors he considered a control for a class of partial differential equations. The later work of Egorov on controls of equations with distributed parameters has been also summarized in the paper of Pontryagin [30].

In [32], and [33] Russell developed Pontryagin's principle for symmetric hyperbolic equations.

Control of equations in Hilbert and Banach space setting has been considered by many authors including Fattorini [13], [14], Blum [3], Egorov [10]. While this setting is more natural for certain problems of control of elastic systems, there are some basic unanswered questions, preventing its use at the present time.

The control of a vibrating string has been considered by Russell in [32]. The basic principles of optimal controls for vibrating beams have been developed by the author in [17], and for vibrating thin plates in [18]. This monograph is mainly devoted to summarizing and unifying these results. Because of the limited scope of this book, many related results will be deliberately ignored. In particular no attempt will be made to relate this work to the parallel developments following the classical techniques of calculus of variations. (For example we shall mention only briefly the very important works of Cesari, Gamkrelidze, Hestenes, or Lurie, which do have close connection with the results presented here).

We shall deliberately ignore many related problems and concentrate specifically on one topic: the optimal control of vibrating elastic system from the point of view of Pontryagin's theory. The simplest cases--the vibrating strings and membranes can be considered in the more general setting of the optimal control of symmetric hyperbolic systems.

In chapter 1 we shall review the basic results of Butkovskii, Lerner and Russell. We shall resist the temptation to consider the related cases, such as the control of electromagnetic wave propagation and restrict the discussion to the control of corresponding elastic systems, i.e. vibrating strings and membranes. Chapter 2 discusses the problem of control of vibrating beams, and mainly repeats the results of the author of [17], [19]. Chapter 3 discusses the same problem for the thin plates from the point of view of [18].

CHAPTER I

A summary of some results on controls of hyperbolic partial differential equations

1.1 The Basic Equations

In this monograph we shall not study the important problem of controllability. The control problems shall be stated and certain sufficiency conditions will be developed for the optimality of a control. The much deeper mathematical questions will be avoided, such as stating the precise conditions characterizing controls of such problems as well posed.

A related question, whether the energy of the system can be reduced to some given value in a finite time will also be avoided. In the specific cases studied in Chapters 2, 3, and 4, this question is answered in the positive. In the very general case posed by equation (1.1) in Chapter I this will remain unanswered. The properties of symmetric hyperbolic systems were first studied by K. O. Friedrichs, and his terminology is adapted here.

We consider a system of partial differential equations of the form

$$L(\underset{\sim}{W}) = E(x)\frac{\partial \underset{\sim}{W}}{\partial t} - A(x)\frac{\partial \underset{\sim}{W}}{\partial x} - C(x)\,\underset{\sim}{W} = \phi(x,t) \;, \qquad (1.1)$$

In this monograph $\phi(x,t)$ is always assumed to be of the form

$$\phi(x,t) = B(x)u(t).$$

$0 \leq t < \infty$, $a \leq x \leq b$, where E, A, and C are $n \times n$ matrices, $\underset{\sim}{W}(x,t)$, $\underset{\sim}{\phi}(x,t)$ are n-dimensional vectors.

An important simplification occurs when the vector function $\underset{\sim}{\phi}(x,t)$ is of the form: $\underset{\sim}{\phi}(x,t) = B(x)\,\underset{\sim}{u}(t)$ where $B(x)$ is an $n \times m$ matrix ($m \leq n$) while $\underset{\sim}{u}(t)$ is an m-dimensional vector. $B(x)$ is given a priori and only the vector $\underset{\sim}{u}(t)$ is allowed to vary over a specified set of admissible controls.

Physically this may mean that the locations at which the control forces are applied are fixed and only the magnitude (and direction) of this force may be varied as functions of time.

The corresponding homogeneous equation is obtained by setting

$$\underset{\sim}{u}(t) \equiv 0, \quad 0 \leq t \leq T$$

$$L_H(W) = E(x)\frac{\partial \underset{\sim}{W}}{\partial t} - A(x)\frac{\partial \underset{\sim}{W}}{\partial x} - C(x)\underset{\sim}{W} = 0 \qquad (1.1^{\underline{H}})$$

In the language of physics the equation $(1.1^{\underline{H}})$ corresponds to a free motion of the system. The function $B(x)u(t)$ is called the forcing function, or the inhomogeneous term. In some problems of control theory physical conditions may necessitate a fixed form of the matrix $B(x)$. The case when only $u(t)$ is to be determined will be considered here. $u(t)$ shall be called the control function, or simply the control.

The equation:

$$L^*(\underset{\sim}{W}) = E(x)\frac{\partial \underset{\sim}{W}^*(x,t)}{\partial t} - \left\{ A(x)\frac{\partial \underset{\sim}{W}^*(x,t)}{\partial x} + \left[\frac{\partial A(x)}{\partial x} - C^T(x) \right] \underset{\sim}{W}^*(x,t) \right\}$$

$$= B(x)\underset{\sim}{u}(t) \qquad (1.1^*)$$

will be called the adjoint equation. The corresponding homogeneous equation is:

$$L^*(W) = 0 \quad , \qquad (1.1^{*H})$$

which is obtained as before as before by putting $\underset{\sim}{u}(t) \equiv 0$ $0 \leq t \leq T$ in the equation (1.1^*). In the case when $L^*(W) = L(W)$, i.e. $C(x) = \frac{\partial A(x)}{\partial x} - C^T(x)$, the equations (1.1), $(1.1^{\underline{H}})$ are called self adjoint.

In each case we can define the norms $|| \; ||_i$, i - 1,2, and we can find a number $M > 0$, such that $||u(t)||_1 \leq M$, and $||B(x)||_2 < M$. The matrices E and A are symmetric, while E is also a stricktly positive matrix on the interval $a \leq x \leq b$. Such system will be called symmetric hyperbolic.

The positive definite property of E implies that all the roots $\lambda_1(x), \lambda_2(x) \ldots \lambda_n(x)$ of the determinantal equation:

$$\det [E(x)\lambda(x) - A(x)] = 0$$

are real valued functions of x. We may assume that on the interval [a,b] we have the inequality: $\lambda_1(x) > \lambda_2(x) > \ldots > \lambda_n(x)$.

The system described by these properties is totally hyperbolic in the terminology of Petrovski or Hörmander.

The solution $\underset{\sim}{W}(x,t)$ is assumed to obey the given initial conditions: $\underset{\sim}{W}(x,0) = \underset{\sim}{W}_0(x)$, where $\underset{\sim}{W}_0(x)$ is an absolutely continuous function on [a,b], whose derivative is uniformly bounded almost everywhere.

We also specify some boundary conditions at the points x = a, x = b, (such as: $\underset{\sim}{W}(a,t) = \underset{\sim}{W}(b,t) = 0$, or $A_a\underset{\sim}{W}(a,t) = 0$, $A_b\underset{\sim}{W}(b,t) = 0$, where A_a, A_b are $n \times n$ constant matrices).

The boundary conditions, which are generally physically motivated, assure the uniqueness of solutions of the differential equation (1.1). With each equation of the form (1.1) we shall associate the corresponding homogeneous equation, obtained by putting $\underset{\sim}{u}(t) \equiv 0$, $0 \leq t \leq T$.

$$E(x)\frac{\partial W_H(x,t)}{\partial t} - A(x)\frac{\partial W_H(x,t)}{\partial x} - C(x)W_H(x,t) = 0 \qquad (1.1^H)$$

Equations of the form (1.1), (1.1^H) are respectively special cases of the more general equations (1.2) and (1.2^H).

$$L(\underset{\sim}{W}) = E(\underset{\sim}{x})\frac{\partial \underset{\sim}{W}(\underset{\sim}{x},t)}{\partial t} - \sum_{i=1}^{n} A_i(\underset{\sim}{x})\frac{\partial \underset{\sim}{W}(\underset{\sim}{x},t)}{\partial x_i} - C(\underset{\sim}{x})\underset{\sim}{W}(\underset{\sim}{x},t)$$

$$= B(\underset{\sim}{x})\underset{\sim}{u}(t)\,, \qquad (1.2)$$

and

$$L(w) = 0 \,. \qquad (1.2^H)$$

Examples of equations of this type are:

<u>The telegraph line equation</u>, which is written as a pair of equations

$$\left.\begin{aligned} L(x)\,\frac{\partial i}{\partial t} + \frac{\partial v}{\partial t} &= f_1(x)u_1(t) \\ C(x)\,\frac{\partial v}{\partial t} + \frac{\partial i}{\partial x} &= f_2(x_1)u_2(t) \end{aligned}\right\} \qquad (a)$$

where i is the current, v the voltage, L, C are respectively inductance and capacitance per unit length.

If we denote by W, E, A respectively:

$$\underset{\sim}{W} = \begin{bmatrix} i \\ v \end{bmatrix} \;, \quad E = \begin{bmatrix} L(x) & 0 \\ 0 & C(x) \end{bmatrix}$$

$$A = \begin{bmatrix} 0 & -1 \\ -1 & 0 \end{bmatrix} \quad B(x) = \begin{bmatrix} f_1(x) & 0 \\ 0 & f_2(x) \end{bmatrix}$$

$$\underset{\sim}{u} = \begin{bmatrix} u_1 \\ u_2 \end{bmatrix}$$

the system of equations (a) becomes:

$$E(x)\frac{\partial W}{\partial t} - A\frac{\partial W}{\partial x} + B(x)\underset{\sim}{u}(t) = 0$$

which is a special case of (1.1). We note that the total energy is given here by:

$$\mathcal{E} = \int_{x=0}^{x=1} (Li^2 + Cv^2)dx$$

<u>Maxwell's equations</u>

These equations for inhomogeneous, anisotropic dielectric medium are:

$$\left.\begin{aligned} \sum_{n=1}^{3} \varepsilon_{jn}(\underset{\sim}{x}) \frac{\partial E_n}{\partial t} - (\nabla \times \underset{\sim}{H})_j &= f_1(x)u_1(t) \\ \sum_{n=1}^{3} \mu_{jn}(\underset{\sim}{x}) \frac{\partial H_n}{\partial t} + (\nabla \times \underset{\sim}{E})_j &= f_2(x)u_2(t) \end{aligned}\right\} \quad (b)$$

again the system (b) is easily put in the form of equation (1.2) by writing:

$$\varepsilon(x) = \begin{bmatrix} \varepsilon_{11}(x) & \varepsilon_{12}(x) & \varepsilon_{13}(x) & 0 & 0 & 0 \\ \varepsilon_{21} & \varepsilon_{22} & \varepsilon_{23} & 0 & 0 & 0 \\ \varepsilon_{31} & \varepsilon_{32} & \varepsilon_{33} & 0 & 0 & 0 \\ 0 & 0 & 0 & \mu_{11} & \mu_{12} & \mu_{13} \\ 0 & 0 & 0 & \mu_{21} & \mu_{22} & \mu_{23} \\ 0 & 0 & 0 & \mu_{31} & \mu_{32} & \mu_{33} \end{bmatrix}$$

$$\underset{\sim}{W} = \begin{bmatrix} E_1 \\ E_2 \\ E_3 \\ H_1 \\ H_2 \\ H_3 \end{bmatrix}$$

and with the operator matrix

$$\sum_{n=1}^{3} A^n \frac{\partial}{\partial x_n} = \begin{bmatrix} 0 & 0 & 0 & 0 & -\frac{\partial}{\partial x_3} & \frac{\partial}{\partial x_2} \\ 0 & 0 & 0 & \frac{\partial}{\partial x_3} & 0 & -\frac{\partial}{\partial x_1} \\ 0 & 0 & 0 & -\frac{\partial}{\partial x_2} & \frac{\partial}{\partial x_1} & 0 \\ 0 & \frac{\partial}{\partial x_3} & -\frac{\partial}{\partial x_2} & 0 & 0 & 0 \\ \frac{\partial}{\partial x_3} & 0 & \frac{\partial}{\partial x_1} & 0 & 0 & 0 \\ \frac{\partial}{\partial x_2} & -\frac{\partial}{\partial x_1} & 0 & 0 & 0 & 0 \end{bmatrix},$$

Maxwell's equations reduce to the form (1.2). The total energy contained in the region Ω is given by:

$$\mathcal{E}(t) = \frac{1}{2} \iiint_{\Omega} \sum_{i,j=1}^{3} (\varepsilon_{ij} E_i E_j + \mu_{ij} H_i H_j)\, dx_1 dx_2 dx_3 \quad .$$

In [40] C. Wilcox gives additional examples of equations of mathematical physics which are of the type (1.2), including the accoustic equation:

$$C(x)\frac{\partial^2 p}{\partial t^2} - \rho(x)\nabla(\frac{1}{\rho}\nabla p) = B(x)u(t)$$

and the classical equations of wave propagation in an anisotropic elastic medium:

$$\rho \frac{\partial^2 W_i}{\partial t^2} - \sum_{i,j,k=1}^{3} \frac{\partial}{\partial x_j}(C_{ijk\ell}\frac{\partial W_k}{\partial x_\ell}) = \sum_j B_{ij}(x)u_j(t)$$

<u>The control problem for the system (1.1)</u>. We shall call $\mathcal{E}(w,t)$ given by:

$$\mathcal{E}(u,t) = \frac{1}{2}\int_a^b [\underset{\sim}{W}(x,t) \cdot E(x,t)\underset{\sim}{W}(x,t)]dx$$

the total energy associated with the control $\underset{\sim}{u}(t)$.

We wish to find an optimal control $\bar{\underset{\sim}{u}}(t)$ such that for some given T > 0, we have:

$$\mathcal{E}(\bar{\underset{\sim}{u}}(t,T) = \min_{u\varepsilon U} (\mathcal{E}(\underset{\sim}{u}(t,T)) .$$

This control problem shall be called the fixed time

interval control problem. (i.e. we have fixed the time interval [0,T] and we seek an admissible control u such that the total energy $\mathcal{E}$ assumes the smallest possible value at the time T.

Definition of the space of admissible controls. The set of controls U will be the class of all functions, or distributions as defined in this paragraph, which will be considered as the inhomogeneous term $B(x)\underset{\sim}{u}(t)$ in the equation (1.1). If B(x) is regarded as fixed we need to specify only the class of distributions (functions) from which $\underset{\sim}{u}(t)$ is to be chosen.

Most assumptions found in mathematical papers concerning the optimum controls of ordinary or partial differential equations include measurability and boundedness of $B(x)\underset{\sim}{u}(t)$. (See for example Butkovskii and Lerner [4], or Russell [32].) In our case, when the time interval [0,T] is finite, these hypothesis would imply square integrability of the inhomogeneous term. Additional assumption of absolute integrability and uniform boundedness in the L_1 norm are commonly included in the hypothesis:

$$\int_a^b |B(x)u(t)|dx \leq M \quad \text{for some } M > 0, \text{ for all } t \in [0,\infty],$$

where M is an a priori given constant. Without any loss of generality this inequality may be replaced by:

$$\int_a^b |B(x)u(t)|dx \leq 1 \quad \text{for all } t \in [0,\infty]. \qquad (N1)$$

In the case when the control has the physical dimension of force this implies that the control force is distributed in a manner determined by a bounded measurable functions B(x), u(t), and that the total force does not exceed unity at any time. Such controls shall be called admissible distributed controls. However, the distributed controls may fail to contain suitable controls necessary for the formulation and solution of important optimal principles. In the problem of optimal control of a vibrating membrane, or of a string, the optimal controls B(x)u(t) (with B(x) not given a priori) turn out to be the point loads (that is the Dirac delta function). Hence the Dirac delta function has to be included among admissible control loads if we want certain control problems to have solutions. In fact most engineers would feel unhappy if point loads and point moments were excluded from the consideration. On physical grounds we may exclude controls $\Phi = B(x)u(t)$ where u(t) is an impluse function (that is again a Dirac delta function, or its derivative), since this would imply the ability of the system to transmit infinite force. In each case the physical considerations must dictate the choice of admissible controls. If generalized functions, or distributions in the sense of Schwartz are allowed as controls, we may again require $||B(x)\underline{u}(t)|| \leq 1$, where the norm may be defined as the usual norm of a functional:

$$||\Phi|| = \sup <\Phi, W> , \; ||W|| = 1 ,$$

over the space of test functions of admissible solutions $\underset{\sim}{W}(x,t)$ of equation (1.1), or of some subclass of such solutions which is to be specified. (It is assumed that the admissible solutions of (1.1) form a normed space.) Since the control $B(x)\underset{\sim}{u}(t)$ may be a distribution, it is clear that the derivatives of $\underset{\sim}{W}(x,t)$ in equation (1.1) mav fail to exist in the classical sense, and are either distributions, or weak derivatives in the sense of Sobolev. (See [35], section 5, pages 39-41 for definition.)* Since every bounded integrable function can also be regarded as a distribution, we can shorten our definitions and subsequent arguments by defining the control $B(x)\underset{\sim}{u}(t)$ to be a distribution whose norm is bounded by unity.

For physical reasons we may prefer to keep the distinction between distributed controls (bounded measurable functions) and point controls (distributions whose support consists of a finite number of points.) We rewrite accordingly for the one dimensional case ($B(x)$ is a real function):

$$B(x) = \phi(x) + \sum_{i=1}^{n} \psi_i(x - \xi_i) \quad ,$$

(*The Sobolev derivative Df of a function f is a function Df such that $\int (Df)\phi = -\int fD\phi$ for every test function ϕ.)

where $\phi(x)$ is an absolutely integrable function on $[a,b]$, whose norm is given by : $||\phi|| = \int|\phi(x)|dx$, and $\psi_i(x - \xi_i)$ is a distribution whose support consists of a single point $\xi_i \in [a,b]$, with $||\psi_i|| = \sup \langle\psi_i, f\rangle$, $f \in L_2 \cap L_1$, $||f||_1 \leq 1$. We stipulate that $B(x)u(t)$ is an admissible control if for some $M > 0$, $||B(x)|| \leq M$. and if for every $\tau \in [0,T]$

$$||B(x)|| \cdot ||u(\tau)|| \leq 1, \quad \text{where}$$

$||B(x)||$ is defined by

$$||B(x)|| = ||\phi|| + \sum_{i=1}^{n} ||\psi_i||.$$

If $B(x)$ is given a priori and only $\underset{\sim}{u}(t)$ is to be chosen, it is frequently true, that the generality of our results is not restricted if $\underset{\sim}{u}(t)$ is chosen from the class of piecewise continuous functions, rather than bounded and measurable functions. Again we impose the restriction $||B(x)u(t)|| \leq 1$, where the norm of the vector $B(x)\underset{\sim}{u}(t)$ can be redefined to suit the particular physical situation. This norm is defined at all points $\tau \in [0,T]$ at which $\underset{\sim}{u}(t)$ is defined. Even if T assumes the value of $+\infty$, there will be at most finitely many points on any finite subinterval of $[0,T]$ on which this norm fails to be defined.

<u>The statement of the main theorem for the fixed interval control problem.</u> We shall consider the symmetric hyperbolic system

$$L(\underset{\sim}{W}) = E(x)\frac{\partial \underset{\sim}{W}}{\partial t} - A(x)\frac{\partial \underset{\sim}{W}}{\partial x} - C(x)\underset{\sim}{W} = B(x)\underset{\sim}{u}(t) \ , \qquad (1.1)$$

$0 \leq t \leq T$, $a \leq x \leq b$, where $E(x)$ is a symmetric, positive definite $n \times n$ matrix, $A(x)$ is a symmetric $n \times n$ matrix, $C(x)$ is an $n \times n$ matrix, $B(x)$ is $n \times m$ matrix, $\underset{\sim}{u}(t)$ is an m-dimensional vector, $m \leq n$, on the domain $D = [0,T] \times [a,b]$.

We assume that the elements of the matrices $A(x)$, $E(x)$, $C(x)$ are continuous functions possessing weak derivatives of order at least two on $[a,b]$.

We pose the following initial value problem:

Let the initial condition be $\underset{\sim}{W}(x,0) = W_o(x)$. (C1)

$W_o(x)$ is absolutely continuous on $[a,b]$. Let the derivative $\frac{d(W_o(x))}{dx}$ exist almost everywhere on $[a,b]$ and be bounded by some constant M (whenever it exists). In addition let the boundary conditions at the points $x = a$, $x = b$ be of the form:

$$\left.\begin{aligned} A_a \underset{\sim}{W}(a,t) &\equiv 0 \\ A_b \underset{\sim}{W}(b,t) &\equiv 0 \end{aligned}\right\} \quad 0 \leq t \leq T \qquad (B1)$$

where A_a, A_b are constant $n \times n$ matrices. (Obviously the conditions:

$$\left.\begin{aligned} A_a W_o(a) &= 0 \\ A_b W_o(b) &= 0 \end{aligned}\right\} \qquad \text{(B2)}$$

must be satisfied.)

The problem posed by equation (1.1) with the initial condition (C1) and the boundary conditions (B1) shall be called the initialboundary value problem (IBVP). A proof of existence of solutions of IBVP is outlined in Courant and Hilbert [44], and a detailed proof for the above case is given in the appendix of [32] by D. Russell. Moreover, if some smoothness properties are assumed for the matrices E, A, C, ($B, A, E, C \in C^2$ is the condition imposed in [32], although it can be easily weakened), then we can show that the IBVP has a unique solution $W(x,t)$ which is continuous in D, absolutely continuous on lines x = constant, and satisfies $W(x,t) = W_H(x,t)$

$$+ \int_a^b\int_0^t G(x-\xi, t-\tau)B(\xi)u(\tau)\,d\tau\,d\xi \quad , \qquad (1.4)$$

where the kernel $G(x-\xi, t-\tau)$ depends upon the boundary conditions (B1), but is independent of either $\phi(x,t) = B(x)\underset{\sim}{u}(t)$, or of the initial conditions (C1).

W_H stands for the solution of the homogeneous equation (1.1 H) with identical initial and boundary conditions.

The formula (1.4) is called Duhamel's principle and shall be used repeatedly in our proofs.

We are now ready to pose a control problem for the IBVP. We shall consider only controls $\underset{\sim}{\phi}(x,t)$ of the form:

$$\underset{\sim}{\phi}(x,t) = B(x)\underset{\sim}{u}(t)$$

with B(x) given a priori, $||B(x)|| = 1$, and $\underset{\sim}{u}(t)$ a bounded measurable function on [0,T], obeying $||u(t)|| \leq 1$.

Controls $\underset{\sim}{u}(t)$ satisfying the above conditions shall be called <u>admissible</u>. The set of all admissible controls will be denoted by U.

<u>Solution of the problem</u>

<u>Theorem 1</u>

Let $\bar{\underset{\sim}{u}}(t)$ be the optimal control for the fixed time interval control problem posed for the initial and boundary value problem (IBVP) (Equations (1.1) and conditions (B1) and (C1)).

Let $\underset{\sim}{W}_H^*$ be the solution of the homogeneous equation of the adjoint system:

$$E(x)\frac{\partial \underset{\sim}{W}_H^*}{\partial t} - A(x)\frac{\partial \underset{\sim}{W}_H^*}{\partial x} - [\frac{\partial A(x)}{\partial x} - C^T(x)]\underset{\sim}{W}_H^* = 0 \qquad (1.1_H^*)$$

satisfying the same boundary conditions as (1.1) and satisfying the same terminal conditions as the optimal solution $\bar{\underset{\sim}{W}}(x,t)$ of (1.1), corresponding to the optimal control u(t);

(i.e., $W_H^*(x,T) = \tilde{W}(x,T)$. Then at every point τ of continuity of $\tilde{u}(t)$, $\tau \in [0,T]$, we have:

$$- \int_a^b [W_H^*(x,t)B(x)]dx \cdot \tilde{u}(\tau) = \max_{u \in U} \{- \int_a^b [W_H^*(x,t)B(x)]dx \cdot \tilde{u}(\tau)\} \tag{1.3}$$

The proof is lengthy, but basically follows the classical arguments of Pontryagin. For the sake of completeness it will be given at the end of this chapter.

Remarks

The statement of the theorem 1 is meaningless unless the following statements are correct:

1) there exists at least one optimal control function $\tilde{u}(t)$;
2) even if this optimal control is not unique (which turns out to be the case in most problems) then the terminal condition $\tilde{W}(x,T)$ resulting from any optimal control should be unique.

These objections are taken care of in the lemmas (1) and (2) stated below.

We introduce the functional $\mathcal{E}(t) = \frac{1}{2} \int_a^b W(x,t) \cdot E(x)W(x,t)dx$ which will be called the total energy of the system, and pose the following control problem:

Among the admissible controls U choose $\tilde{u} \in U$ such that the corresponding solution $W(\tilde{u}(t),x,t)$ possesses the smallest total energy at the time t = T, i.e.

$$\tilde{\mathcal{E}}(T) = \frac{1}{2}\int_a^b \tilde{W}(\tilde{u},x,T)\cdot E(x)\tilde{W}(\tilde{u},x,T)dx$$

$$\leq \frac{1}{2}\int_a^b W(u,x,T)\cdot E(x)W(u,x,T)dx,$$

for any $u \in U$. This control problem will be called the fixed ([0,T]) interval control problem.

In our discussion we shall make use of the following inner product:

$$\langle W_1, W_2\rangle = \int_a^b [W_1(x,t)\; E(x)W_2(x,t)]dx = f(t) \quad (1.5)$$

where $W_1(x,t)$, $W_2(x,t)$ are arbitrary L_2 vectors for which the above Riemann integral exists. The operation $\cdot$ is performed as usual: $\phi \cdot \psi = \phi_1\psi_1 + \phi_2\psi_2 + \ldots + \phi_n\psi_n$.

In the following discussion $W_1(x,t)$, $W_2(x,t)$ shall be always the solutions of equation (1.1) or of the equation (1.1^H), as stated. It is an elementary exercise to show that the product $\langle W_1, W_2\rangle$ as given by formula (1.5) satisfies all axiomatic postulates of an inner product. The norm generated by the inner product is:

$$||W||_E = \sqrt{<W,W>} = \sqrt{2\,\mathcal{E}(W)} \quad ,$$

$$\mathcal{E}(W) = \frac{1}{2}||W||_E^2 \quad .$$

It will be called the energy norm of $W(x,t)$.

Since $<W_1,W_2>$ is an inner product, the Cauchy-Schwartz inequality is correct:

$$<W_1,W_2>^2 \leq ||W_1||_E^2 \; ||W_2||_E^2 = \frac{1}{4}\,\mathcal{E}(W_1)\,\mathcal{E}(W_2). \qquad (1.6)$$

We shall now state a lemma which turns out to be very useful in the proof of theoreml.

Lemma 1

Let $W(x,t)$ be a solution of IBVP corresponding to an admissible control $u(t)$, $t \in [0,T]$, and W_H^* be the solution of the homogeneous adjoint equation $(1.1H^*)$, where both W and W_H^* are continuous in $D = [0,T] \times [a,b]$ and absolutely continuous on lines $x =$ constant, with $\frac{\partial W}{\partial t}$, $\frac{\partial W_H^*}{\partial t}$ being of the class L_2 in D, (and defined almost everywhere). Then we have the relationship:

$$\frac{d}{dt} <W,W_H^*> = \int_a^b [W_H^*(x,t)B(x)]dx\cdot u(t)$$

almost everywhere in D.

Note: Quadratic forms are written here as $yA\cdot z$ (y, z-k, ℓ vectors respectively, A-$\ell\times k$ matrix). This will always mean that y is an k-row vector, z - ℓ - column vector. The entries are always real valued functions, and the transpose signs are deliberately supressed.

Proof

$$\langle \underset{\sim}{W}(x,t),\ \underset{\sim}{W}^*_H(x,T)\rangle - \langle \underset{\sim}{W}(x,0),\underset{\sim}{W}^*_H(x,0)\rangle$$

$$= \int_a^b \underset{\sim}{W}^*_H(x,T)E(x)\underset{\sim}{W}(x,T)\,dx - \int_a^b \underset{\sim}{W}^*_H(x,0)E(x)\underset{\sim}{W}(x,0)\,dx$$

$$= \int_a^b \{ \int_0^T \frac{d}{dt} (\frac{\partial \underset{\sim}{W}^*_H}{\partial t} (x,t)\ E(x)\underset{\sim}{W}(x,t)$$

$$+ \underset{\sim}{W}^*_H(x,t)E(x)\frac{\partial \underset{\sim}{W}(x,t)}{\partial t})dt\}\ dx\ .$$

The last step was justified by the hypothesis that $\underset{\sim}{W}^*_H$ and $\underset{\sim}{W}$ are absolutely continuous on lines x = constant.

We can now use Fubini's theorem to obtain:

$$\langle \underset{\sim}{W}^*_H(x,T),\underset{\sim}{W}(x,T)\rangle - \langle \underset{\sim}{W}^*_H(x,0),\underset{\sim}{W}(x,0)\rangle$$

$$= \int_0^T \int_a^b \{ \frac{\partial \underset{\sim}{W}^*_H(x,t)}{\partial t} E(x)\underset{\sim}{W}(x,t) + \underset{\sim}{W}^*_H(x,t)E(x)\frac{\partial \underset{\sim}{W}(x,t)}{\partial t} \}\ dxdt$$

Using the fact that $E(x)$ is symmetric (and therefore

$$\frac{\partial \underset{\sim}{W}^*_H(x,t)}{\partial t} E(x)\underset{\sim}{W}(x,t) = \underset{\sim}{W}(x,t)E(x)\frac{\partial \underset{\sim}{W}^*_H(x,t)}{\partial t}),$$

and substituting equations (1.1) and (1.1^{*H}) we have:

$$\langle W_H^*(x,T), W(x,T)\rangle - \langle W_H^*(x,0), W(x,0)\rangle$$

$$= \int_0^T \int_a^b \{W(x,T)[A(x)\frac{\partial W_H^*(x,t)}{\partial x} + (\frac{\partial A(x)}{\partial x} - C^T(x))W_H^*(x,t)]$$

$$+ W_H^*(x,t)[A(x)\frac{\partial W_H^*(x,t)}{\partial x} + C(x)W_H^*(x,t) + B(x)u(t)]\}\, dxdt$$

$$= \int_a^b \{ \int_0^T (W(x,t)A(x) \frac{\partial W_H^*(x,t)}{\partial x} + W(x,t)\frac{\partial A(x)}{\partial x} W_H^*(x,t)$$

$$+ W_H^*(x,t)A(x)\frac{\partial W(x,t)}{\partial x} + W_H^*(x,t)B(x)u(t))dx\}\, dt$$

(Fubini's theorem was used again to effect the change in the order of integration.)

Our hypothesis allow integration by parts of the term:

$$W(x,t)A(x)\frac{\partial W_H^*(x,t)}{\partial x} \quad \text{on the interval } a \le x \le b,$$

and we obtain:

$$\langle W(x,T), W_H^*(x,T)\rangle - \langle W(x,0), W_H^*(x,0)\rangle$$

$$= \int_0^T \{W(b,t)A(b)W_H^*(b,t) - W(a,t)A(a)W_H^*(a,t) - \int_a^b (W_H^*(x,t)A(x)$$

$$\frac{\partial W(x,t)}{\partial t} + W_H^*(x,t)\frac{\partial A(x)}{\partial x}W(x,t))dx + \int_a^b (W_H^*(x,t)A(x)\frac{\partial W(x,t)}{\partial x}$$

$$+ \underset{\sim}{W}(x,t)\frac{\partial A(x)}{\partial x}\, \underset{\sim}{W}_H^*(x,t) + \underset{\sim}{W}_H^*(x,t)B(x)\underset{\sim}{u}(t))dx)dt$$

The boundary conditions (B1) imply the vanishing all terms except the last, and we have:

$$\langle \underset{\sim}{W}(x,T),\, \underset{\sim}{W}_H^*(x,T)\rangle - \langle W(x,0), W_H^*(x,0)\rangle$$

$$= \int_0^T \{ \int_a^b \underset{\sim}{W}_H^*(x,t)B(x)\underset{\sim}{u}(t)dx \}\, dt \qquad (1.3)$$

We observe that the limits 0,T played no part in this argument, and that for arbitrary $t_1, t_2 \in [0,T]$ we can obtain a similar result:

$$\langle W(x,t_1), W_H^*(x,t_1)\rangle - \langle W(x,t_2), W_H^*(x,t_2)\rangle$$

$$= \int_{t_1}^{t_2} \{ \int_a^b W_H^*(x,t)B(x)u(t)dx\}dt$$

It follows that for measurable controls we have:

$$\frac{d}{dt} \langle W(x,t), W_H^*(x,t)\rangle = \int_a^b W_H^*(x,t)B(x)u(t)dx$$

almost everywhere on [0,T], proving the lemma.

<u>Corollary</u>

A self adjoint system (1.1^H) is conservative relative

to the energy product $\mathcal{E}(W(x,t),t)$.

Proof

Putting $u(t) \equiv 0$, we have:

$$\frac{d}{dt} \langle \underset{\sim}{W}_H, \underset{\sim}{W}_H^* \rangle = 0.$$

If the system is selfadjoint $\underset{\sim}{W}_H^* \equiv \underset{\sim}{W}_H$, and

$$\frac{d}{dt} \langle W_H, W_H \rangle = 0, \ t \in [0,T] ,$$

implying

$$\langle \underset{\sim}{W}_H, \underset{\sim}{W}_H \rangle = \text{constant}.$$

By definition $\frac{1}{2} \langle \underset{\sim}{W}, \underset{\sim}{W} \rangle$ is the energy $\mathcal{E}(\underset{\sim}{W}(x,t),t)$. Hence, $\mathcal{E}(\underset{\sim}{W}(x,t),t) = \mathcal{E}(\underset{\sim}{W}(x,0),0)$ = constant, that is the energy $\mathcal{E}(\underset{\sim}{W},t)$ of the system is conserved.

Notation

We shall denote by $L_{2(m)}(I)$ the space consisting of all m-dimensional vectors $\phi(t), t \in I$, such that $\int_I ||\phi||^2 dt < \infty$, where $|| \ ||$ denotes the usual Euclidean norm of an m-vector, that is: $||\phi||^2 = \phi_1^2 + \phi_2^2 + \ldots + \phi_m^2$. As usual we consider only the equivalence classes of such vectors that is ϕ and ϕ' are identified with each other if their components differ only on a set of measure zero in I.

In our discussion I will stand for the interval [0,T].

<u>Lemma 2</u> (The basic convexity lemma)

If $\underset{\sim}{u}_1(t)$ and $\underset{\sim}{u}_2(t)$ are admissible displacements, then $\underset{\sim}{u} = \lambda \underset{\sim}{u}_1 + (1 - \lambda)\underset{\sim}{u}_2$ is also an admissible displacement for any $0 \leq \lambda \leq 1$.

The proof is almost trivial. If $\underset{\sim}{u}_1$, $\underset{\sim}{u}_2$ are bounded uniformly, and measurable, then so is $\underset{\sim}{u}$. It only needs to be shown that $||\underset{\sim}{u}|| \leq 1$. We have the triangular inequality:

$$||\underset{\sim}{u}|| \leq \lambda||\underset{\sim}{u}_1|| + (1 - \lambda)||\underset{\sim}{u}_2|| \leq \lambda + (1 - \lambda) \leq 1,$$

as was to be proved. It follows that the set of admissible controls U is convex.

It is also trivial to show that U is closed. Hence, U is a weakly compact, convex subset of $L_{2\,(m)}(0,T)$. We note that the above result (the convexity lemma) would carry over to the case when more general controls $\Phi(x,t)$ are considered where $\Phi(x,t)$ may be a generalized function (distribution) of x for a fixed t, and a bounded measurable function of t for fixed x.

The existence of an optimal control is proved in

<u>Lemma 3</u>

Let the conditions on A(x), B(x), C(x), E(x) be such that the existence theorem quoted above (and discussed in [40]) is true; that is the IBVP does have a solution $\underset{\sim}{W}(\underset{\sim}{u},x,t)$ for each admissible control $\underset{\sim}{u}(t)$, such that $||W(\underset{\sim}{u},x,t)||$ is a continuous, uniformly bounded function of t on [0,T], and the bound M may be chosen independently of $\underset{\sim}{u}(t)$. Then there exists a control $\underset{\sim}{u}(t) \in U$, such that $\mathcal{J}(\tilde{\underset{\sim}{u}}(t),T) \geq \mathcal{J}(\underset{\sim}{u}(t),T)$ for any $\underset{\sim}{u} \in U$.

<u>Proof</u>: Clearly there exists a greatest lower bound K on all real numbers $\mathcal{J}(u(t),T) \geq 0$, $u \in U$, and a sequence of controls $u_i(t)$ $t \in [0,T)$ such that for the corresponding solutions $W_i(u_i,x,t)$ we have

$$\lim_{i\to\infty} \mathcal{J}(W_i(x,t),T) = K .$$

Since $u \in U$ implies $||u|| \leq 1$ the sequence $\{u_i(t)\}$ has a subsequence $\{u_k\}$ in $L_2(I)$ which converges weakly to some element $\tilde{u}(t) \in L_2(I)$. It is trivial to show that $\tilde{u}(t) \in U$. All that remains to be shown is $\mathcal{J}(\tilde{u},T) = K$, and therefore that $\tilde{u}(t)$ is an optimal control. We use the Duhamel's principle (1.4):

$$\tilde{W}(x,t) = W_H(x,t) + \int_a^b \int_0^t G(x-\xi,t-\tau)B(\xi)\tilde{u}(\tau)d\xi d\tau$$

$$W_H(x,t) + \int_0^t G'(x,t-\tau)u(\tau)d\tau,$$

We denote for each k = 1,2,...

$$W_k(x,t) = W_H(x,t) + \int_0^t G'(x,t-\tau)u_k(\tau)d\tau.$$

Since $\underset{\sim}{u}_i(\tau)$ converges weakly to $\tilde{\underset{\sim}{u}}(\tau)$ we obtain the result:

$$\lim_{k\to\infty} \mathcal{E}(W_k - \tilde{W}) = 0 , \quad \tilde{w} = \tilde{w}(\tilde{u}) ,$$

and

$$\lim_{k\to\infty} (W_k, T) = \lim_{k\to\infty} \frac{1}{2} \int_a^b W_k(x,T)E(x)W_k(x,T)dx$$

$$= \frac{1}{2} \int_a^b \tilde{W}(x,T)E(x)\tilde{W}(x,T)dx$$

$$= \mathcal{E}(\tilde{W},T)$$

$$= K$$

which was to be proved.

Remark

Some easy examples can be given here to demonstrate that the optimal control is in general not unique. An obvious example is one where the non-uniqueness is so bad that we shall stipulate later that this type of problem is to be excluded from our discussion.

Consider the specific physical problem of a vibrating string, with some initial data given $W(x,0) = W_0(x)$, and with the boundary conditions $W(a,t) = W(b,t) \equiv 0$, for all $t \in [0,T)$. It may turn out that an application of an admissible control is capable of reducing the total energy $\mathcal{E}$ to the zero level at some time $\tau < T$. Then any control which does produce $\mathcal{E}(T) = 0$ is an optimal control according to our definition. The problem of control of the vibrating string shall be discussed in Chapter 2, at which time it should become clear why I labelled this type of non-uniqueness "very bad". But even if $\delta(T) > 0$ for any admissible control, there still may exist more than one optimal control reducing the total energy of the string to its lowest possible value at the time T. This type of non-uniqueness does not prevent us from formulating Pontryagin-type principles for optimal controls, and it is assumed to be present in almost all of our problems.

Definition

Let $\underset{\sim}{u}(t)$ be an admissible control, and $W(\underset{\sim}{u},x,t)$ be the corresponding solution of the IBVP. If $\tilde{u}(t)$ is an optimal (admissible) control, then $\tilde{W}(\tilde{\underset{\sim}{u}},x,t)$ shall be called the optimal solution of IBVP.

Lemma 4

The sets of optimal controls, and of optimal solutions are convex. That is if $\tilde{\underset{\sim}{u}}_1,\tilde{\underset{\sim}{u}}_2$ are optimal controls then $\tilde{\underset{\sim}{u}} = \lambda\tilde{\underset{\sim}{u}}_1 + (1-\lambda)\tilde{\underset{\sim}{u}}_2$, $0 \leq \lambda \leq 1$, is also an optimal control, and $\tilde{W} = \lambda W_1(u_1,x,t) + (1-\lambda)W_2(u_2,x,t) = \tilde{W}(\tilde{u},x,t)$ is the corresponding optimal solution.

Proof

Let us first show that the solution $\tilde{W}(\tilde{u})$ does obey the linear formula: $\tilde{W}(\tilde{u}) = \lambda W_1(u_1) + (1-\lambda)W_2(u_2)$, whenever

$\tilde{u} = \lambda u_1 + (1-\lambda)u_2$. We apply Duhamel's principle:

$$\lambda W_1 + (1-\lambda)W_2 = \lambda[W_H + \int_0^t G(x,t-\tau)u_1(\tau)d\tau]$$

$$+ (1-\lambda)[W_H + \int_0^t G(x,t-\tau)u_2(\tau)d\tau]$$

$$= W_H + \int_0^t G(x,t-\tau)\tilde{u}(\tau)d\tau$$

$$= \tilde{W} \quad ,$$

as required.

Now let us assume that $\tilde{u}_1(t)$, $\tilde{u}_2(t)$ are both optimal controls, that is, $\mathcal{E}(\tilde{u}_1,T) = \mathcal{E}(\tilde{u}_2,T) = E = \min_{u \in U} \mathcal{E}(u,T)$.

$$\mathcal{E}(\lambda u_1 + (1-\lambda)u_2,t) = \mathcal{E}(\tilde{u},t) = \frac{1}{2}||\tilde{W}||_E^2$$

$$= \frac{1}{2} <\tilde{\underset{\sim}{W}},\tilde{\underset{\sim}{W}}>$$

$$= \frac{1}{2} <\lambda \underset{\sim}{W}_1 + (1-\lambda)\underset{\sim}{W}_2 \ ,$$

$$\lambda \underset{\sim}{W}_1 + (1-\lambda)\underset{\sim}{W}_2 = \frac{1}{2}\{\lambda^2<\underset{\sim}{W}_1,\underset{\sim}{W}_1> + (1-2\lambda+\lambda^2)$$

$$<\underset{\sim}{W}_2,\underset{\sim}{W}_2> + 2\lambda(1-\lambda)<\underset{\sim}{W}_1,\underset{\sim}{W}_2>\}$$

In particular, if we put $t = T$, we obtain:

$$\mathcal{E}(\tilde{u},T) = \frac{1}{2}[2\lambda^2 E + 2(1-2\lambda + \lambda^2)E + 2(\lambda-\lambda^2)\langle W_1, W_2\rangle_{t=T}]$$

Since $\lambda \geq \lambda^2$ $(0 \leq \lambda \leq 1)$, and since by Cauchy-Schwartz inequality, we have:

$$\langle W_1, W_2\rangle_{t=T} \leq \sqrt{\langle W_1, W_1\rangle_T \langle W_2, W_2\rangle_T} = \sqrt{4E \cdot E} = 2E,$$

we obtain:

$$\mathcal{E}(\tilde{u},T) \leq (1-2\lambda + 2\lambda^2)E + 2(\lambda-\lambda^2)E = E.$$

This shows that $\tilde{u}$ is again an optimal control, completing the proof. In fact we have proved more than we claimed. Since E was the lowest energy attainable, and $\mathcal{E}(\tilde{u},T) \leq E$, we must have the strict equality $\mathcal{E}(\tilde{u},T) = E$. Retracting our arguments we see that this is possible only if $\langle W_1, W_2\rangle^2_{t=T} = \langle W_1, W_1\rangle_{t=T} \cdot \langle W_2, W_2\rangle_{t=T}$, that is if the Cauchy-Schwartz inequality is a strict equality. This in turn implies that for all x in the interval [a,b], $W_1(x,T) = \alpha W_2(x,T)$, for some constant α. (The statement "a·e." in [a,b] is replaced by "everywhere" because of continuity.) However, $\langle W_1, W_2\rangle = 2E$ implies that $\alpha = 1$, and we must have: $W_1(x,T) = W_2(x,T)$. Since u_1, u_2 were arbitrary optimal controls, we have obtained a uniqueness lemma for the finite value of optimal solutions of IBVP.

Lemma 5

Let $u_1(t)$, $u_2(t)$ be optimal controls for the fixed interval control problem, $t \in [0,T]$ and $W_1(x,t)$, $W_2(x,t)$ the corresponding optimal solutions of IBVP. Then we must have $\underset{\sim}{W}_1(x,T) = \underset{\sim}{W}_2(x,T)$ everywhere on the interval: $a \leq x \leq b$. ($\underset{\sim}{W}$ stands as before for the vector

$$\underset{\sim}{W} = \begin{pmatrix} W \\ \frac{\partial W}{\partial t} \end{pmatrix} .)$$

Corollary

If the set of admissible controls is such that to each finite value of the solution corresponds at most one control, then the optimal control is unique.

Remark

It is easy to show that the set of attainable solutions is also convex, bounded and closed in the topology of $L_{2(m)}[a,b]$.

Proof of the maximum principle (Theorem 1)

The proof follows exactly the arguments of Pontryagin et. al. in [28]. We shall make a more restricted statement, that is, we seek a piece-wise continuous control $\underset{\sim}{u}(t)$, rather than measurable and bounded one, as originally postulated in [28]. Because of this the proof is greatly simplified. The extension of the proof to bounded measurable functions can then be handled as a straight forward corollary. Let $\tau \in (0,T)$ be a point of continuity for the optimal control

$\tilde{u}(t)$, and I_δ be an interval such that u(t) is continuous on the open interval $I_\delta = (\tau-\delta,\tau+\delta)$. We choose δ sufficiently small so that $I_\delta \subset [0,T]$. We modify the optimal control $\tilde{u}(t)$ as follows:

$$\hat{u}_{\varepsilon,\delta}(t) = \tilde{u}(t) \text{ if } t \in [0,T] - I_\delta$$

$$= \dot{u}(t) + \varepsilon\eta(t) \text{ if } t \in I_\delta ,$$

with $\eta(t)$ an arbitrary chosen function and the constant ε chosen: $|\varepsilon| \leq \varepsilon_o$, with $\varepsilon_o(\eta(t)) > 0$ selected so that $\tilde{u} + \varepsilon\eta$ is an admissible control whenever $|\varepsilon| < \varepsilon_o$. If no $\varepsilon_o > 0$ is available, such that for some $\eta(t)$ ($t \in I_\delta$), $\check{u} + \varepsilon\eta \in U$ for all ε $|\varepsilon| < \varepsilon_o$, we can reduce the size of I_δ. If for all δ, the set of ε_o, such that $u + \varepsilon\eta \in U$, $|\varepsilon| < \varepsilon_o$ is empty in I_δ, the maximum principle is trivially correct in the interval I_δ, and there is nothing to prove. We assume therefore that there exists an $\varepsilon_o > 0$, such that $\hat{u}_{\varepsilon,\delta}(t) = \tilde{u} + \varepsilon\eta$ is an admissible control in I_δ, for all ε, such that $|\varepsilon| < \varepsilon_o$.

Since $\tilde{u}$ was optimal, we must have

$$\mathcal{E}(\tilde{u},T) \leq \mathcal{E}(u_{\varepsilon,\delta},T) \text{ but by linearity (see 1.4)}$$

$\hat{W}(\hat{u}_{\varepsilon,\delta},x,t) = \tilde{W}(\tilde{u},x,t) + \varepsilon W_\eta(\eta,x,t)$, so that

$$\mathcal{S}(u_{\varepsilon,\delta},T) = \mathcal{S}(\tilde{u},x,T) + \varepsilon\langle W_\eta,\tilde{W}\rangle_{t=T} + \varepsilon^2\mathcal{S}(W_\eta,x,T) \qquad \text{(b)}$$

Combining the inequalities (a) and (b) we see that

$$\varepsilon\langle W_\eta,\tilde{W}\rangle_{t=T} + \varepsilon^2\mathcal{S}(W_\eta,T) \geq 0 \text{ for all } \varepsilon,\ |\varepsilon| < \varepsilon_0,$$

or $\varepsilon^2\mathcal{S}(W_\eta,T) \geq -\varepsilon\langle W_\eta,\tilde{W}\rangle$ which implies that for $\varepsilon > 0$
$\varepsilon\,\mathcal{S}(W_\eta,T) \leq \langle W_\eta,\tilde{W}\rangle$ and for $\varepsilon < 0$

$$-\varepsilon\,\mathcal{S}(W_\eta,T) \leq \langle W_\eta,\tilde{W}\rangle.$$

Combining these inequalities we have:

$$|\varepsilon|\,\mathcal{S}(W_\eta,T) \leq \langle W_\eta,\tilde{W}\rangle \text{ which implies:}$$

$$\langle W_\eta,\tilde{W}\rangle_{t=T} \geq 0 \text{ for all } \varepsilon,\ \varepsilon_0 > |\varepsilon| \geq 0\ . \qquad \text{(c)}$$

We recollect that $W_\eta(x,t)$ was the solution corresponding to the control $u_\eta(t) = \begin{cases} 0, & t \notin I_\delta \\ \eta(t), & t \in I_\delta\ , \end{cases}$

so that on the interval $[\tau+\delta,T]$ W_η is a solution of a homogeneous equation (1.1^H), while $W_\eta \equiv 0$ on $[0,\tau-\delta]$. Hence if W_H^* is a solution of the homogeneous adjoint equation for IBVP; we have according to the corollary to lemma 1,

$$\frac{d}{dt} \langle W_H^*, W_\eta \rangle = 0 \text{ for all } t \in [\tau+\delta, T] \qquad (d)$$

If W_H^* satisfies the same final conditions as $\tilde{W}$, we have

$$\langle W_H^*, W_\eta \rangle_{t=T} = \langle W_\eta, \tilde{W} \rangle_{t=T} \geq 0 \text{ by inequality (c).}$$

Combining this with (d), we obtain:

$$\langle W_H^*, W_\eta \rangle \geq 0, \text{ for all } t \in [\tau+\delta, T].$$

On the interval $[0,\tau-\delta] W_\eta \equiv 0$, and $\langle W_H^*, W_\eta \rangle \equiv 0$, for all $t \in [0,\tau-\delta]$. We only need to show that for a sufficiently small interval $[\tau-\delta,\tau+\delta]$ the product $\langle W_H^*, W_\eta \rangle$ does not become negative. We use the formula:

$$\frac{d}{dt} \langle W_H^*, W_\eta \rangle = \{\int_a^b W_H^*(x,t)B(x)dx\}. \; u_{\varepsilon,\delta}(t),$$

almost everywhere; so that using (d) we obtain:

$$\langle W_H^*, W_\eta \rangle_{t=\tau+\delta} = \int_{t-\delta}^{t+\delta} \{\int_a^b W_H^*(x,t)B(x)\hat{u}_{\varepsilon,\delta}(t)dx\}dt \geq 0, \qquad (e)$$

By assumed properties of $B(x)$, W_H^*, $u(t)$ it follows that

$$\lim_{\delta\to 0} \frac{1}{2\delta} \int_{\tau-\delta}^{\tau+\delta} W_H^* B(\hat{u}_{\varepsilon,\delta} - \tilde{u})dt = 0$$

uniformly for all ε, $0 < |\varepsilon| \leq \varepsilon_o$, and

$$\lim_{\delta \to 0} \int_{\tau-\delta}^{\tau+\delta} \frac{1}{2\delta} \int_a^b [W_H^*(x,t_1) - W_H^*(x,t_2)]$$

$$B(x)\tilde{u}(t)]dt = 0, \quad t_1,t_2 \in I_\delta.$$

Together with (e) this implies that

$$\int_a^b (W_H^*(x,t)\cdot B(x)dx)[\hat{u}_{\varepsilon,\delta}(t) - \tilde{u}(t)] \geq 0 \quad \text{in } I_\delta.$$

Since

$$<W_H^*,W_\eta> = \int_0^t \int_a^b W_H^*(x,t)B(x)u_\eta(t)dxdt,$$

we have

$$\int_a^b W_H^*(x,t)B(x)[u_{\varepsilon,\delta}(t) - \tilde{u}(t)]dx \geq 0 \quad \text{on } [0,T] - I_\delta \text{ by (c)}.$$

Therefore

$$\int_a^b W_H^*(x,t)B(x)u_{\varepsilon,\delta}(t)dx \geq \int_a^b W_H^*(x,t)B(x)\tilde{u}(t)dx \quad \text{for all } t \in [0,T].$$

This is the required form of the maximum principle, completing the proof of theorem 1.

This pattern of proof is almost identical (apart from a few simplifications) with the one offered by Russell in [32], and generally parallels the arguments used by Pontryagin in [28].

Some remarks concerning theorem 1

Let us first repeat again the definition of $W_H^*(x,t)$. $W_H^*(x,t)$ is the solution of the adjoint homogeneous equation $(1.1\underline{H}^*)$ which assumes the same final condition as $\tilde{W}(x,t)$, that is, as the optimal solution of (1.1); i.e. $W_H^*(x,T) = \tilde{W}(x,T)$. At first glance it appears that the maximum principle as stated by the theorem 1 is hard to apply in the general case, and is entirely useless if $\delta(\tilde{u},T) = 0$, since then $\tilde{W}(x,T) = 0$, and $W_H^*(x,T) = 0$, implying $W_H^*(x,t) \equiv 0$ for all t.

If this difficulty is avoided, that is if $\delta(\tilde{u},T) > 0$, then frequently the maximum principle does convey a useful information, and in some cases uniquely determines the optimal control. Examples will be given in the discussion of optimal controls for beams and plates where the maximum principle can be used to demonstrate that a given control is not an optimal control and to effect improvements in arbitrary selected controls.

Let us look again at the integral:

$$\int_a^b \underset{\sim}{W}_H^*(x,t)B(x)\cdot\tilde{\underset{\sim}{u}}(t)dx, \quad \tilde{\underset{\sim}{u}} \in U,$$

where U is the set of all admissible controls, that is, of measurable controls on $t \in [0,T]$, with $||\underset{\sim}{u}||_m \leq 1$, for all $u \in U$. Let us assume that the norm $||\ ||_m$ is defined as follows:

$$||\underset{\sim}{u}||^2_{(m)} = u_1^2 + u_2^2 + \ldots + u_m^2.$$

That is, U is a unit sphere in the m-dimensional Euclidean space. Then

$$\int_a^b \underset{\sim}{W}_H^*(x,t)B(x)\cdot\underset{\sim}{u}(t)dx$$

can assume a maximum only if $\tilde{\underset{\sim}{u}}(t)$ is a unit vector parallel to the vector $\int_a^b \underset{\sim}{W}_H^*(x,t)B(x)dx$, that is, if:

$$\tilde{\underset{\sim}{u}}(t) = \frac{\int_a^b \underset{\sim}{W}_H^*(x,t)B(x)dx}{||\int_a^b W_H^*(x,t)B(x)dx||_{(m)}} \tag{1.6}$$

If m = 1, we obtain the well-known version of the "bang bang" optimal control:

$$\tilde{u}(t) = \text{sign}\,\{\int_a^b W_H^*(x,t)B(x)dx\}. \tag{1.6^a}$$

Since the final condition $W_H^*(x,T)$ is unique, the solution of homogeneous boundary value problem $W_H(x,t)$, $t \in [0,T]$ is also unique, and in this case the optimal control $\tilde{u}(t)$ is uniquely determined by (1.6) or (1.6^a). This is not necessarily true if a different definition is given for the set of admissible controls U. Let us for example choose the following definition of U: $u \in U$ if $u(t)$ is piecewise continuous, and $||u||_{m,\infty} \leq 1$, where $||\ ||_{m,\infty}$ denotes:

$$||u||_{m,\infty} = \max \{ |u_1|, |u_2|, \ldots, |u_m| \}.$$

In this situation it is clear that the optimum control will be in general nonunique.

Take any vector $\tilde{\underset{\sim}{u}}(t)$ parallel to $\underset{\sim}{\phi}(t) = \int_a^b \underset{\sim}{W}_H^*(x,t)B(x)dx$ and of maximum norm, i.e. $||\tilde{u}||_\infty = 1$. Take any nonzero vector $\underset{\sim}{v}(t)$ which is orthogonal to $\underset{\sim}{\phi}(t)$, and such that $||\tilde{u}||_\infty = ||\underset{\sim}{u} + \underset{\sim}{v}||_\infty = 1$, (which is possible in this case). Now, if

$$\int_a^b W_H(x,t)B(x)\cdot\tilde{u}(t)dx = \max_{u \in U} \int_a^b W_H(x,t)B(x)\cdot u(t),$$

that is if $\tilde{\underset{\sim}{u}}(t)$ obeys the maximum principle, then so does the control $(\tilde{\underset{\sim}{u}} + \underset{\sim}{v})$.

We note that this nonuniqueness disappears when $m = 1$,

so in the one dimensional case we obtain again the unique control given by the formula (1.6[a]). It is not hard to see that optimal control is unique if the definition of the norm $|| \; ||$ is such that U forms a unit ball which is strickly convex, (i.e. if its boundary contains no straight line intervals). Conversely, if the unit ball is not strictly convex in the norm $|| \; ||_m$ then there will exist non-unique optimal controls. The construction of such a case is analogous to the example given above for the sup. norm. (This does not mean that it is not possible to have a unique control, if the unit ball fails to be strictly convex.)

A more subtle question is: If the solutions are non-unique how badly non-unique are they? "How badly" could be defined by a number

$$k = \sup \int_0^T ||\bar{u} - \bar{v}||dt, \text{ where the sup. is taken}$$

over all optimal solution $\tilde{u}$, $\tilde{v} \in U$. If the solutions are unique, how close are they to being non-unique? "How close" can be defined by numbers η_ε, as follows. Choose $\varepsilon > 0$ and, U_ε the subset of U of all controls such that if $u_\varepsilon \in U_\varepsilon$, then $|\mathcal{J}(u_\varepsilon,T)_T - \mathcal{J}(\tilde{u},T)| < \varepsilon$.

Define $\eta_\varepsilon = \sup\limits_{U_\varepsilon} \int_0^T ||\check{u} - u_\varepsilon||dt$. Define $\eta = \lim\limits_{\varepsilon \to 0} (\frac{\eta_\varepsilon}{\varepsilon})$, and if the limit does not exist put $\eta = \infty$. In this we conjecture that the answers to these questions are closely related to the modulus of convexity of the unit ball $\delta_U(\varepsilon)$ of U.

$$\delta_U(\varepsilon) = \frac{1}{2}\,\text{inf.}\ (2 - ||x + y||),\ ||x|| = ||y|| = 1,$$

$$||x - y|| > \varepsilon.$$

<u>An example of an optimal control of a symmetric hyperbolic system</u>

We consider the one dimensional wave propagation (say a string), governed by the equation:

$$\rho\,\frac{\partial^2 W}{\partial t^2} - \tau\,\frac{\partial^2 W}{\partial x^2} = \phi(x)u(t)\ , \qquad (1.8)$$

where ρ, τ are given nonzero constants, $\left\{\begin{array}{l} x \in [0,1] \\ t \in [0,T] \end{array}\right.$,

with the boundary conditions:

$$\left.\begin{array}{l} W(0,t) \equiv 0 \\ W(1,t) \equiv 0 \end{array}\right\} \qquad (1.8^B)$$

$\phi(x)$ is given, while $u(t)$ is to be determined so that at the time $t = T$ the total energy $\mathcal{E}(T)$ is to assume the lowest possible value. We assume the following:

1) For a fixed $x \in [0,1]$ $W(x,t)$ is a continuously differentiable function of t, and $\frac{\partial W}{\partial t}$ is uniformly bounded on $D = [0,1] \times [0,T]$. $\frac{\partial^2 W}{\partial t^2}$ is defined almost everywhere on D, and is a continuous function of x and a piece-wise continuous function of t.

2) For a fixed $t \in [0,T]$ the solution $W(x,t)$ is a bounded absolutely continuous function of x on the interval $0 \leq x \leq 1$, and possess weak derivatives of order at least two in D. (See the definition of Sobolev [35], section 5, pages 39-41).

3) $\phi(x)$, $0 \leq x \leq 1$, is a sum a bounded measurable function, and of the distribution:

$$\phi(x) = \sum_{i=1}^{k} c_i \delta(x - \xi_i)$$

where δ is the Dirac delta function, and ξ_i, $i = 1,2,\ldots,k$ are points chosen in the interior of the interval $[0,1]$.

4) $u(t)$, $t \in [0,T]$ is a piecewise continuous function, obeying the restriction: $|u(t)| \leq 1$ for all $t \in [0,T]$.

Note: Clearly these conditions are not independent of each other and one could show that some of the properties 3), 4) are implied by 1) and 2) or one could start with 3) and 4) and derive the basic properties of the solutions of (1.8) and (1.8^B). We note that $\frac{\partial u}{\partial t}$, $\frac{\partial u}{\partial x}$ are square integrable on [0,1]. We shall consider the simplest case when $k = 1$ in condition 3), that is:

$$\phi(x) = \delta(x - \xi), \quad \xi \in (0,1).$$

Our control shall consist of a point load applied at the point $x = \xi \in (0,1)$.

Before we discuss the solution of this problem, we should confirm the statement in the title of this section, that is that the system we are discussing is a symmetric hyperbolic system and that it can be reduced to the equation (1.1) with the matrix E positive definite symmetric, and A symmetric. We write

$$W_1(x,t) = \frac{\partial W(x,t)}{\partial t}$$

$$W_2(x,t) = \frac{\partial W(x,t)}{\partial x}$$

$\underset{\sim}{W}$ will denote the vector:

$$\underset{\sim}{W}(x,t) = \begin{Bmatrix} W_1(x,t) \\ W_2(x,t) \end{Bmatrix}$$

The equation (1.8) can now be rewritten in the standard form:

$$E \frac{\partial}{\partial t} \underset{\sim}{W} + A \frac{\partial}{\partial x} \underset{\sim}{W} = B(x)\underset{\sim}{u}(t) \tag{1.9}$$

where

$$E = \begin{bmatrix} \frac{1}{c^2} & 0 \\ 0 & 1 \end{bmatrix}, \quad c^2 = \frac{\tau}{\rho}$$

$$A = \begin{bmatrix} -1 & 0 \\ 0 & -1 \end{bmatrix}$$

$$B(x) = \begin{bmatrix} \frac{\phi(x)}{\tau} & 0 \\ 0 & 0 \end{bmatrix}$$

$$\underset{\sim}{u} = \begin{bmatrix} u(t) \\ 0 \end{bmatrix}$$

The system (1.9) is selfadjoint.

The total energy of the system is given by:

$$\mathcal{E}(W,t) = \frac{1}{2} \int_0^1 W(x,t)E \cdot W(x,t)\,dx$$

$$= \frac{1}{2\tau} \int_0^1 [\rho (\frac{\partial W}{\partial t})^2 + \tau (\frac{\partial W}{\partial x})^2]dx$$

By the corollary to lemma 1 the total energy is conserved if $u(t) \equiv 0$.

We are now ready to solve the above posed problem of optimal control for this system. Since the system is self-adjoint $W_H = W_H^*$, and the optimal control $\bar{u}(t)$ must satisfy the relationship (1.3)

$$\int_0^1 (\underset{\sim}{W}_H(x,t)B(x)dx\cdot \bar{\underset{\sim}{u}}(t) = \min_{u\in U} \int_0^1 \underset{\sim}{W}_H(x,t)B(x)dx\cdot \underset{\sim}{u}(t),$$

$$\underset{\sim}{W}_H(x,t)B(x) = \begin{bmatrix} \frac{1}{\tau}\, \frac{\partial W_H(x,t)}{\partial t}\cdot \delta(x-\xi) \\ \\ 0 \end{bmatrix}$$

$$\int_0^1 \underset{\sim}{W}_H B(x)dx = \begin{bmatrix} \frac{1}{\tau}\, \frac{\partial W_H(\xi,t)}{\partial t} \\ \\ 0 \end{bmatrix}$$

and

$$\int_0^1 \underset{\sim}{W}_H B(x)dx\cdot \underset{\sim}{u}(t) = \begin{bmatrix} \frac{1}{\tau}\, \frac{\partial W_H(\xi,t)}{\partial t}\cdot u(t) \\ \\ 0 \end{bmatrix} .$$

Ignoring the zero term in this vector equation (that is ignoring the equation 0 = 0), we have:

$$\frac{1}{\tau}\frac{\partial W_H(\xi,t)}{\partial t}\tilde{u}(t) = \min_{u\varepsilon U}\frac{1}{\tau}\frac{\partial W_H(\xi,t)}{\partial t}u(t)$$

This problem has the solution:

$$\tilde{u}(t) = -\text{ sign}\left[\frac{\partial W_H(\xi,t)}{\partial t}\right]$$

where the function sign (f) is defined by:

$$\text{sign}(f) = \begin{cases} -1 & \text{if} \quad f < 0 \\ 0 & \text{if} \quad f = 0 \\ 1 & \text{if} \quad f > 0 \end{cases}$$

We repeat again that $W_H(x,t)$ is the solution of the homogeneous equation (1.8) with the property:

$$\tilde{W}(\tilde{u}(t),x,T) = W_H(x,T),$$

where $\tilde{W}$ is the optimum solution, corresponding to the optimum control $\tilde{u}(t)$. We conclude that any optimal control will be constant on intervals of length $\theta/2$, where θ is the period during which the system vibrating freely will complete one cycle. On such time intervals the optimal control will alternately assume the values of +1 and -1.

An entirely different control problem could be posed if a

control of a vibrating string (1.8) is effected by means of a prescribed displacement at a single point. In [33] Russell discusses such optimal control at the boundary point $x = b$,

$$\hat{W}(b,t) = \hat{u}(t)$$
$$\hat{W}(a,t) \equiv 0$$

and on the open interval (a,b) the displacement obeys the equation

$$\rho \frac{\partial^2 \hat{W}}{\partial t^2} - \tau \frac{\partial^2 \hat{W}}{\partial x^2} = 0 \quad . \qquad (1.8^a)$$

$\hat{u}(t)$ is assumed to obey the constraint:

$$-1 \leq \frac{\partial^2 \hat{u}}{\partial t^2} \leq +1 \ .$$

Upon a change of the variable $W(x,t) = \hat{W}(x,t) + xu(t)$ the problem is reduced to the familiar form

$$\frac{\partial^2 W}{\partial t^2} - c^2 \frac{\partial^2 W}{\partial x^2} = xu(t),$$

$$u(t) = - \frac{\partial^2 \hat{u}}{\partial t^2}$$

$$u(a,t) \equiv 0$$

$$u(b,t) \equiv 0$$

$$|u| \leq 1$$

and the optimal control satisfying the relationship (1.3) is

$$\overset{\vee}{u}(t) = \text{sign} \int_a^b [x \cdot \frac{\partial W(x,t)}{\partial t}] \, dx.$$

Other problems of optimal control for a vibrating string, could be easily suggested. Perhaps the most important comment to be made at this point is upon the limitation of our results. We have carefully avoided the problem of allowing the matrix $B(x)$ to be also optimized, or even more generally to consider an arbitrary control of the form $\Phi(x,t)$ as the right hand side of the equation (1.8).

In some physical cases this may not be a significant limitation, since a control servomechanism is naturally positioned at some fixed position in space and the force exerted by it is to be optimized as a function of time only.

In cases where the position itself is to be optimized additional theoretical difficulties arise, since in general the optimal control turns out to be a generalized function of the space variables, whose magnitude and position as a function of time may be of the "chattering" type in the terminology of the Russian school. (See [43] page 289 for a heuristic explanation).

A closer look at this phenomenon will be attempted in the

more detailed discussion of the control problem for a vibrating plate.

Control of the equations of the type (2.1)

We comment that no changes are necessary in all proofs of lemmas and theorems of this section if the variable x is replaced by an n-dimensional vector $\underset{\sim}{x}$ whose domain is the parallelopiped $a_i \leq x_i \leq b_i$, $i = 1,2,\ldots,n$.

All proofs were given for the one dimensional case only for the sake of simplicity of argument. It is easily checked that all arguments are valid if a,b, are replaced by the corresponding vectors $\underset{\sim}{a}$, $\underset{\sim}{b}$, and x is replaced by $\underset{\sim}{x}$.

Remark

Connection between the results listed in this chapter and the boundary control principles for a vibrating string, discussed in an article of Butkovskii and Poltavski [8] is not immediately apparent. Despite the similarity of postulation of the control problem, Butkovskii and Poltavski use a different definition of the space of admissible controls. The basic equation is again

$$\frac{\partial^2 W}{\partial t^2} - \frac{\partial^2 W}{\partial x^2} = 0.$$

The control u(t) is applied at the boundary point x = 0 and

has to satisfy the constraint

$$[\int_0^T |u(t)|^q dt]^{1/q} \leq 1, \quad 1 \leq q < \infty.$$

By the application of the method of moments the authors obtain the following formula for the optimal control $\bar{u}(t)$, which reduces the total energy to zero level in the shortest time:

$$\bar{u}(t) = \begin{cases} \frac{1}{\pi(n+1)} (F(t) + C), & t \in [0,\tau] ; \\ \frac{1}{\pi n} (F(t) + C), & t \in [\tau,T] , \end{cases}$$

if τ, $2\pi > \tau > 0$, is given $\tau = 2\pi n - T$. Here $F(t)$ is a certain function defined on the whole real axis with a period of 2π, and C is a constant. (See [8]).

An instantly optimal control

Definitions and some basic theorems

The main difficulty which arises in the use of Pontryagin's principle arises from the fact that the principle is postulated in terms of the finite state of the system. A possible way of avoiding this difficulty is to redefine what is meant by the "optimal control". In fact one definition of "optimal" may be suggested by the technique of optimizing which any engineer would intuitively use to reduce the energy of the system. It turns out that this is exactly the definition which allows us to postulate the maximality principle in terms of the initial conditions, or in terms of the current state of the system, rather than of the finite state

We shall now introduce a limiting process, which will allow us to resolve some of the difficulties. We shall first need the following theorem.

Theorem 4. Let $\tilde{\phi}(x,t)$ be a time optimal control reducing the total energy of the system to the value $E < \mathcal{E}(0)$ in the shortest possible time $T > 0$. Then $\tilde{\phi}(x,t)$ is an optimal control for the fixed interval $[0,T]$ control problem

Proof. Assume to the contrary that $\tilde{\phi}(x,t)$ is not optimal for the interval $[0,T]$. Hence there exists an admissible control $\psi(x,t)$, such that $\mathcal{E}(\psi(x,t),T) < E$. Since $\mathcal{E}(\psi(x,t),0) = \mathcal{E}(0) > E$, and since $\mathcal{E}(\psi(x,t),t)$ is a continuous function of time, therefore there must be a time $t = \tau$, $0 < \tau < T$, such that $\mathcal{E}(\psi(x,t),0) = E$. This contradicts the time optimal property of the control $\tilde{\phi}(x,t)$, thereby proving the theorem.

Remark 1. The converse theorem is clearly false, as can be demonstrated by examples similar to the one illustrated in Fig. 1.

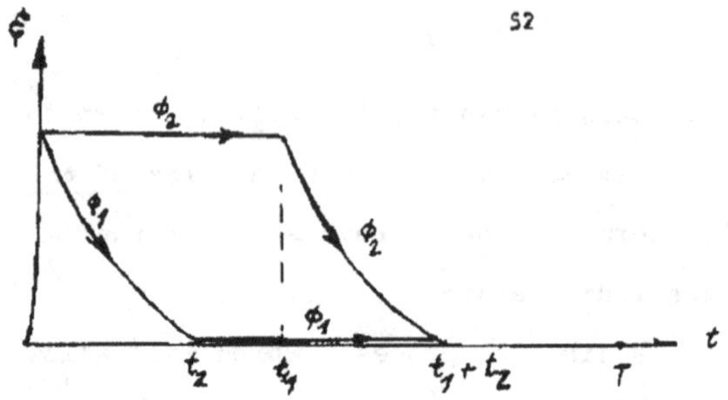

Fig. 1

<u>Definition of instantly optimal control</u>. Let $\phi_0(x)$ be the time optimal control associated with the energy level E, $E < \mathcal{E}(t=0)$, reducing $\mathcal{E}(t)$ to value E in the time $t = T_0$. Theorem 4 asserts that $\tilde{\phi}(x)$ is also an optimal interval control for the interval $[0,T_0]$.

We can subdivide the time interval selecting points t_0, $t_{1,0}$, $t_{2,0}, \ldots, t_{n,0}$; $t_0 = 0$, $t_{n,0} = T_0$, such that the corresponding energy levels are obeying strict inequalities $\mathcal{E}(0) = E_0 > E_{1,0} > E_{2,0} > \ldots > E_{n,0} = E$.

We now introduce a collection of admissible controls $\phi_{1,j}$, such that $\phi_{1,1}$ reduces the energy from E_0 to $E_{1,0}$ in the shortest possible time, say $t_{1,1}$, subject to initial conditions on $w(0,x)$, $\partial w(0,x)/\partial t$ given in the statement of the problem. By lemma 5 the final condition for this subinterval: $\underset{\sim}{w}(t_{1,1}) = [w(t_{1,1}), \partial w(t_{1,1})/\partial t]$ is then uniquely determined by $\phi_{1,1}$. The admissible control $\phi_{2,1}$ reduces the energy to the value $E_{2,1}$ on another time interval $[t_{1,1}, t_{1,2}]$ in the shortest possible time, with the initial conditions for $\phi_{2,1}$ coinciding with the final conditions of $\underset{\sim}{w}(\phi_{1,1})$ at $t_{1,2}$, i. e. with $\underset{\sim}{w}(t_{1,2})$. On each interval we formulate the minimal time control problem and find a solution $\phi_{i,1}$. (These are not necessarily unique.) The collection of controls $\phi_{i,1}$ determines a control ϕ_1 on the time interval $[t_o,T_1]$, $t_o = 0$, $T_1 = t_{1,n}$, which reduces the total evergy to the value E in some time T_1, $T_1 \geq T_o$. Now we subdivide further the energy interval $[E_o,E]$, that is, we subdivide each of the subintervals $[E_{i,1}, E_{i+1,1}]$, and establish in analogous manner a new control $\phi_2(x,t)$. We carry out a sequence of subdivisions, such that

$$\lim_{j\to\infty} (E_{i-1,j} - E_{i,j}) = 0 \quad \text{and} \quad |t_{i-1,j} - t_{i,j}| \to 0.$$

The corresponding controls ϕ_j satisfy the conditions (iii), and we conclude that there exists a unique generalized function $\hat{\phi}(x,t)$:

$$\hat{\phi}(x,t) = \lim_{j\to\infty} \phi_j(x,t), \quad ||\hat{\phi}|| \leq \max_j ||\phi_j||.$$

From the construction of the generalized function $\hat{\phi}(x,t)$ we can immediately deduce some important properties.

<u>Theorem 2.</u> $\hat{\phi}(x,t)$ satisfies the maximum principle:

$$\int_{-\ell/2}^{+\ell/2} [-\hat{\phi}(x,t)\,\frac{\partial w(\hat{\phi},x,t)}{\partial t}]dx = \max \int_{-\ell/2}^{+\ell/2} [-\phi(x,t)\,\frac{\partial w(\phi,x,t)}{\partial t}]dx$$

for all admissible controls $\phi(x,t)$, and for any t in

the domain of the control $\phi(x,t)$, such that $E(t) > E$.

Proof. Given $\varepsilon > 0$ there exists an index K_1, such that on each subinterval $[t_{i-1,j}, t_{i,j}]$,

$$\int_{-\ell/2}^{+\ell/2} [-\phi(x,t)\frac{\partial w(\phi(x,t))}{\partial t} + \phi_j(x,t)\frac{\partial w(\phi_j,x,t_{i,j})}{\partial t}]dx < \frac{\varepsilon}{2}$$

for all $j > K_1$.

Also there exists an index K_2 such that

$$\int_{-\ell/2}^{+\ell/2} [-\phi_j(x,t)\frac{\partial w(\phi_j(x,t),t_{i,j})}{\partial t} + \phi_j \frac{\partial w_H^*(x,t_{i,j})}{\partial t}]dx$$

$$= \int_{-\ell/2}^{+\ell/2} \phi_j(x,t)\cdot\frac{\partial}{\partial t}\{w_H^*(x,t_{i,j}) - w(x,t_{i,j})\}dx < \frac{\varepsilon}{2}$$

for all $j > K_2$ (where $w_H^*(x,t)$ denotes as before the solution of the homogeneous equation satisfying unique final (optimal) condition on $[t_{i,j},t_{i+1,j}]$). This follows from the uniqueness of the finite state:

$$w_H^*(x,t_{i,j}) = w(t_{i+1,j}), \text{ and also}$$

from the fact that w_H^* and w are continuously differentiable, and their time derivatives are uniformly bounded.

Choosing $K = \max(K_1,K_2)$, we have for all $j > K$,

$$\int_{-\ell/2}^{+\ell/2} [-\phi(x,t)\frac{\partial w(\phi(x,t))}{\partial t} + \phi_j \frac{\partial (w_H^*(x,t))}{\partial t}]\, dx < \varepsilon$$

for all $t \in [0,T]$, where $T = \lim_{i\to\infty} T_i$. (In general we have to allow for the possibility that $T = +\infty$, although in the case of a vibrating string it can be shown that T always exists.) Since for each j-th subdivision, $\phi_j(x,t)$ is the optimal control on each fixed interval $[t_{i-1,j}, t_{i,j}]$,

$$\int_{-\ell/2}^{+\ell/2} [-\hat{\phi}_j \frac{\partial w_H(x,t_{i,j})}{\partial t}]dx = \max \int_{-\ell/2}^{+\ell/2} [-\phi(x,t)\frac{\partial w_H(x,t_{i,j})}{\partial t}]dx$$

for an admissible control $\phi(x,t)$ having the same final state $w(x,t_{i,j}) = w_H(x,t_{i,j})$. An obvious argument leads now to the desired conclusion:

$$\int_{-\ell/2}^{+\ell/2} [-\hat{\phi}(x,t)\frac{\partial w(\hat{\phi}(x,t))}{\partial t}]dx = \max \int_{-\ell/2}^{+\ell/2} -[\phi(x,t)\frac{\partial w(\phi(x,t))}{\partial t}]dx.$$

<u>Note:</u> Since this integral is equal to

$$\frac{d}{dt}\langle w(\hat{\phi},x,t), w(\hat{\phi},x,t)\rangle = \frac{d}{dt}\mathcal{E}(w(\hat{\phi},x,t)),$$

the control $\hat{\phi}(x,t)$ has the property that the energy decreases at maximum rate at each point of the time interval $[0,\hat{T}]$.

The principle is valid if $\mathcal{E}(w(\phi(x,t)))>0$ for all

$t \in [0,\hat{T}]$, regardless of whether $\mathcal{E}(T)$ is positive, or equal to zero. The control $\hat{\phi}(x,t)$ shall be called <u>instantly optimal</u>.

<u>Theorem 3.</u> The instantly optimal control $\hat{\phi}(x,t)$ is unique, i.e., $\hat{\phi}(x,t)$ does not depend on the limiting process chosen, or on the properties of the elements of the sequence of time minimal functions $\{\hat{\phi}_i(x,t)\}$.

<u>Proof.</u> Assume that the generalized functions $\hat{\phi}_1(x,t)$, $\hat{\phi}_2(x,t)$, both satisfy $\lim_{i\to\infty} \tilde{\phi}_i^{(1)} = \hat{\phi}_1$ and $\lim_{i\to\infty} \tilde{\phi}_i^{(2)} = \hat{\phi}_2$ for two limiting processes as described above. We use the convexity of the time optimal controls, and continuity to obtain

$$\lim_{i\to\infty} \frac{1}{2}(\tilde{\phi}_i^{(1)} + \tilde{\phi}_i^{(2)}) = \frac{1}{2}(\hat{\phi}_1 + \hat{\phi}_2),$$

which is also an instantly optimal control. If $\underset{\sim}{w}_1(x,t) \triangleq [w(x,t), \partial w(x,t)/\partial t]$ is the displacement vector corresponding to $\hat{\phi}_1(x,t)$ and $w_2(x,t)$ corresponds to $\hat{\phi}_2(x,t)$, then $\frac{1}{2}(w_1 + w_2)$ corresponds to $\frac{1}{2}(\hat{\phi}_1 + \hat{\phi}_2)$ by linearity. By construction of $\hat{\phi}_1$ and $\hat{\phi}_2$ and using the Lemma 5, we have

$$\mathcal{E}(\underset{\sim}{w}_1(\hat{\phi}_1(x,t),t)) = \mathcal{E}(\underset{\sim}{w}_2(\hat{\phi}_2(x,t),t))$$

for all $t \in [0,\hat{T}]$. Hence,

$$\langle w_1, w_1 \rangle = \langle w_2, w_2 \rangle = \frac{1}{4} \langle (w_1 + w_2), (w_1 + w_2) \rangle.$$

the use of Cauchy-Schwartz inequality shows that we have in fact equality $\langle w_1, w_2 \rangle^2 = \langle w_1, w_1 \rangle \cdot \langle w_2, w_2 \rangle$, and the equality $w_1 = w_2$ easily follows for all $t \in [0,T]$. But this implies that $\hat{\phi}_1(x,t)$ and $\hat{\phi}_2(x,t)$ are equal in the sense of our definition, that is, they belong to the same equivalence class.

Remark. It is easy to see that, given $\tilde{E}$, $0 < \tilde{E} < \mathcal{E}(0)$, and given the instantly optimal control $\hat{\phi}(x,t)$, such that $\mathcal{E}(w(\hat{\phi}(x,t),T)) = \tilde{E}$, the time T is either greater or equal to the minimal time $\hat{T}$ corresponding to the minimal time optimum control $\tilde{\phi}(x,t)$. It is important to exhibit cases when a control $\hat{\phi}(x,t)$ exists which is instantly optimal, but is not time optimal, to justify the separate definitions. The possibility of such a situation can be demonstrated by a simple example. Such example is given later in the section 2.9 of chapter 2.

APPENDIX TO CHAPTER I

0.1 Introductory remarks concerning some concepts from the theory of generalized functions. In this monograph the controls $\phi(x,t) = B(x)u(t)$ are either integrable and square integrable (vector) functions on the interval $[-\ell/2, +\ell/2]$ for a fixed t, or $\phi(x,t)$ could be generalized functions (for example, the Dirac delta function) whose supports are a finite number of points in $[-\ell/2, +\ell/2]$. By a theorem due to Lidskii, $B(x)$ is a derivative of a continuous matrix (see [15, p.146]). In our discussion an element of $B(x)$ will be either the Dirac delta function, or its derivative, or else it will be a regular function in $L_2 \cap L_1[-\ell/2, +\ell/2]$. The space of test functions $w(x,t)$, $\partial w(x,t)/\partial t$ can be enlarged to include functions only of the class C^1 in $[-\ell/2, +\ell/2]$, since for all such function the product $(\delta(x - \xi), w(x))$ and $(\delta(x - \xi), \partial w(x,t)/\partial t)$ is defined (for any x in the interval $[-\ell/2, +\ell/2]$). If $\phi_1(x,t)$ is a regular function in $L_2[-\ell/2, +\ell/2])$, then the product (ϕ_1, w) is given by:

$\int_{-\ell/2}^{+\ell/2} \phi_1(x,t)\cdot w(x,t)dx$. We interpret similarly the product $(\phi_2(x,t), \partial w(x,t)/\partial t)$ and $(\underset{\sim}{\phi}, \underset{\sim}{w}) = (\phi_1, w) + (\phi_2, \frac{\partial w}{\partial t})$. Since $w(x,t)$ is integrable in $[-\ell/2, +\ell/2]$, we can define

$$||w(x,t)|| = \int_{-\ell/2}^{+\ell/2} |w(x,t)|dx_{(t=\text{constant})}.$$

The norm of the functional $\phi(x,t)$ can be defined as

$$||\phi|| = \sup_{||w||=1} (\phi,w)$$

$||\phi|| \le K$ will define admissibility of a control. We shall say that a sequence of generalized functions ϕ_i converges to a generalized function ϕ if, for every test function ψ, we have $\lim_{i\to\infty} (\phi_i,\psi) = (\phi,\psi)$. We shall use the accepted notation

$\int_a^b \phi(x,t)\psi(x,t)dx$ instead of $(\phi,\psi)_{[a,b]}$ even though ϕ may not be a logically integrable function.

It is not hard to establish some sufficient conditions for the existence of a unique generalized function ϕ, such that $\lim_{i\to\infty} (\phi_i,\psi) = (\phi,\psi)$ for every test function ψ. Examples of sufficient conditions are given below in (i), (ii), and (iii).

(i) ϕ_i are regular functionals (that is, are locally summable) with a bounded support and obey the hypothesis of the Lebesgue theorem on dominated convergence.

(ii) ϕ_i have a bounded support, decrease (increase) monotonically in every neighborhood of any point which lies in their support, and are bounded below (above) by a locally summable function.

(iii) ϕ_i have a bounded support, their norm is bounded by some constant, and the sequence of numbers (ϕ_i,ψ) converges for every test function ψ.

For the case when it is possible to assign a norm to each test function, the proof is given, as previously stated, in the Lemma in Appendix 1 of [17]. For the proof of this statement, in the case of the test function space K ($\psi \in K$ implies $\psi \in C_0^\infty$) see [15, Appendix A, pp. 368-369].

CHAPTER II

The Optimal Control of Vibrating Beams

2.0. In this chapter we discuss the problem of a transversely vibrating beam, and of a coupled transverse vibration. The optimal control problems are defined and corresponding Pontryagin's principle is developed. The maximum principle is also developed for the instantly optimal control. Some specific examples are discussed in detail.

2.1. The basic equation and the related hypothesis for a transversely vibrating beam

Under the usual simplifying assumption the equation of transverse vibration of a beam is

$$(2.1) \qquad L(w) = \rho(x)\cdot A(x)\cdot\frac{\partial^2 w(x,t)}{\partial t^2} + \frac{\partial^2}{\partial x^2}\left(E(x)\cdot I(x)\cdot\frac{\partial^2 w(x,t)}{\partial x^2}\right) = f(x,t),$$

$$-\ell/2 \le x \le +\ell/2, \quad t \ge 0.$$

The corresponding homogeneous equation is

(2.1a) $$L(w) = 0.$$

(See, for example, [36] for the discussion of the hypothesis of the "simple theory" of bending of beams.)

The physical meaning of the symbols is given below.

$\rho(x)$: the material density;

$A(x)$: the cross-sectional area;

$w(x,t)$: the transverse displacement;

$E(x)$: Young's modulus; (here it is assumed to be constant)

$I(x)$: the moment of inertia of the cross-sectional area about the neutral axis;

$f(x,t)$: the applied load.

Because of the physical meaning of (1) we must assume that the displacement $w(x,t)$ is a continuously differentiable function of x and t. On the interval $[-\ell/2,+\ell/2]$, ρ, A, E, I are uniformly bounded, piecewise smooth, positive functions of x. An additional assumption of material homogeneity would imply that ρ and E are constant. ($A(x)$, $I(x)$ may still vary along the length of the beam.) We assume the correctness of Hooke's law and equate the strain energy with the complementary energy of the beam.

In the remainder of this paper we shall consider only the class of weak solutions of (1) obeying the following conditions:

(a) $w(x,t)$ is a continuously differentiable function of x and of t on $\Omega = [-\ell/2,+\ell/2] \times [0,T]$ (i.e., $\partial w(x,t)/\partial t$ and

$\partial w(x,t)/\partial x$ are continuous functions of x and t on Ω);

(b) $\sqrt{\rho(x)A(x)}\,(\partial w(x,t)/\partial t)$, and $\sqrt{E(x)I(x)}\,(\partial^2 w(x,t)/\partial x^2)$ are square integrable functions of x on $[-\ell/2,+\ell/2]$, and the energy functions

$$K(t) = \frac{1}{2}\int_{-\ell/2}^{+\ell/2} \rho(x)A(x)\left[\frac{\partial w(x,t)}{\partial t}\right]^2 dx, \tag{2.2}$$

$$V(t) = \frac{1}{2}\int_{-\ell/2}^{+\ell/2} E(x)I(x)\left[\frac{\partial^2 w(x,t)}{\partial x^2}\right]^2 dx \tag{2.3}$$

are continuous functions of t, uniformly bounded on the interval $[0,T)$; ($T = +\infty$ may be considered).

(c) $w(x,t)$ obeys one of the three conditions (2.4a), (2.4b), or (2.4c) listed below at each boundary point $x = -\ell/2$, $x = +\ell/2$:

$$\left\{\begin{aligned} w(\pm\ell/2,t) &= 0, \\ \frac{\partial w(\pm\ell/2,t)}{\partial x} &= 0 \end{aligned}\right. \tag{2.4a}$$

(a built-in end);

$$(2.4b)\quad \begin{cases} w(\pm \ell/2,t) = 0, \\ EI\,\dfrac{\partial^2 w(\pm \ell/2,t)}{\partial x^2} = 0 \end{cases}$$

(a freely supported end);

$$(2.4c)\quad \begin{cases} EI\,\dfrac{\partial^2 w(\pm \ell/2,t)}{\partial x^2} = 0, \\ \dfrac{\partial}{\partial x}\left[EI\,\dfrac{\partial^2 w(\pm \ell/2,t)}{\partial x^2}\right] = 0 \end{cases}$$

(a free end);

(d) $w(x,t)$ obeys given initial conditions of the form

$$(2.5a)\qquad w(x,0) = \psi(x),$$

$$(2.5b)\qquad \frac{\partial w(x,0)}{\partial t} = \eta(x),$$

$$\psi(x), \eta(x) \in C^1[-\ell/2, +\ell/2].$$

We also assume that the inhomogeneous term $f(x,t)$ satisfies either of the conditions:

(2,i) $f(x,t)$ is a square integrable (hence absolutely integrable) function of the variable x in the interval

$[-\ell/2,+\ell/2]$, and $\int_{-\ell/2}^{+\ell/2} |f(x,t)|dx$ is a measurable and uniformly bounded function of t in the interval [0,T]. Also f(x,t) is assumed to obey the inequality

$$(2.6) \qquad ||f(x,t)||^2_{(x)} = \left[\int_{-\ell/2}^{+\ell/2} |f(x,t)|dx \right]^2 \leq 1.$$

<u>Note.</u> There is no additional generality in assuming that $||f(x,t)||^2_{(x)} \leq C$ for some $C > 0$.

These conditions imply that the total energy of the beam $\mathcal{E}(t) = K(t) + V(t)$ is also uniformly bounded for all $t \in [0,T]$, where K(t) and V(t) are defined by (2.2) and (2.3) respectively.

The case (2.i) will be called the case of a distributed load, and the control function f(x,t) satisfying the condition (2.i) will be called a distributed load control.

(2.ii) f(x,t) is assumed to be of the form

$$(2.7) \qquad f(x,t) = \sum_{i=1}^{N} \delta(x-\xi_i(t))\phi_i(t) + \sigma(x,t),$$

where $\sigma(x,t)$ is a distributed load control; $\delta(x - \xi_i(t))$ is the "shifted" Dirac delta function, regarded as a generalized function (see [15, pp. 3,4]). The functions $\phi_i(t)$ are measurable functions of the variable t, obeying the condition

$$(2.7a) \qquad \sum_{i=1}^{N} |\phi_i(t)| + \int_{-\ell/2}^{+\ell/2} |\sigma(x,t)|dx \leq 1$$

for all $t \in [0,T]$.

The total energy of the beam is uniformly bounded above by some constant. (We could, of course, incorporate point couples in expression (7a):

$$(2.7b) \quad f(x,t) = \sum_{i=1}^{N} \delta(x-\xi_i(t))\phi_i(t) + \sum_{i=1}^{M} \delta'(x-\zeta_i(t))\mu_i(t)+\sigma(x,t),$$

where ' denotes $\partial/\partial x$.)

For the sake of simplicity no point couples will be used in this work, except as possible limits of sequences considered in the last paragraphs of this chapter.

The functions $\xi_i(t)$ are measurable functions whose domain includes the interval $[0,T]$ and whose range lies in the interval $[-\ell/2,+\ell/2]$. In case (2.ii) we shall denote by $||f(x,t)||_{(x)}$ the quantity

$$||f(x,t)||_{(x)} = \sum|\phi_i(t)| + \int_{-\ell/2}^{+\ell/2} |\sigma(x,t)|dx.$$

As before we assume $||f(x,t)||_{(x)} \leq 1$. The first term on the left-hand side of 2.7[a] will be called the point load controls. A generalized function $f(x,t)$ obeying either of the conditions (2,i) or (2,ii) will be called an admissible control.

2.2 Remarks. The control functions are regarded as generalized functions over the space of test functions satisfying the conditions (a), (b), (c) and (d). It is clear that the products $(\delta(x),w(x,t))(\delta(x),\partial w(x,t)/\partial t)$ are defined for any test function satisfying the condition (a).

The problem of existence and uniqueness of solutions of the mixed boundary and initial value problem (MBVP) posed by (2.1) with conditions (2.4a,b) and (2.5) will not be considered in this paper. In the case when the control function $f(x,t)$ is both an absolutely integrable and a square integrable function of both x and t, proofs can be found in the literature. In the more usual case (2,ii) the author has not been able to find a published proof. However, it is easy to check that the classical proofs can be extended to cover the case when $f(x,t)$ obeys (2.7) by the use of suitable delta-convergent sequences (see [15], col.1, sec. 2.5] for explanation of the procedure). The existence and uniqueness of solutions of the MBVP will be assumed in the subsequent discussion. The important problem of controllability will also be neglected in this monograph.

For purposes of convenience we shall denote the usual

bilinear products $(\phi(x), f(x))$ arising in the generalized function theory by the symbol $\int \phi(x) f(x) dx$ even though $\phi(x)$ may be generalized function which is not locally integrable. (This follows the usual practice in physics and engineering, and avoids the inconvenience of double notation, and of discussion of separate cases.) The solution of the inhomogeneous equation (2.1) subject to conditions (2.4) and (2.5), with a control function $\phi(x,t)$ obeying the conditions (2.i), is known to obey Duhamel's principle

$$(2.8) \quad w(x,t) = w_H(x,t) + \int_0^t \int_{-\ell/2}^{+\ell/2} G(x,\xi,t,\tau)\cdot\phi(\xi,\tau)\,d\xi\,d\tau,$$

where $w_H(x,t)$ is the solution of the homogeneous equation, while $G(x,\xi,t,\tau)$ depends only on the coefficients ρ, A, E, and I, and on the boundary conditions (4), but does not depend on either $\phi(x,t)$ or on the initial value functions $\psi(x)$ and $\eta(x)$.

Again an elementary argument concerning delta-convergent sequences shows that this statement may be extended to cover the case (2,ii). We observe that the admissible controls form a convex set, i.e., if $f_1(x,t)$ and $f_2(x,t)$ are admissible controls, then $\Lambda f_1 + (1 - \Lambda) f_2$ is also an admissible control for any $0 \leq \Lambda \leq 1$.

2.3 The Energy Terms. The kinetic energy of the beam is given by the formula

$$(2.9)\qquad K = \frac{1}{2}\int_{-\ell/2}^{+\ell/2} \rho(x)A(x)\left(\frac{\partial w(x,t)}{\partial t}\right)^2 dx,$$

and the strain energy by

$$(2.10)\qquad V = \frac{1}{2}\int_{-\ell/2}^{+\ell/2} E(x)I(x)\left(\frac{\partial^2 w(x,t)}{\partial x^2}\right)^2 dx.$$

The total energy $\mathcal{E}(t)$ is the sum of the kinetic energy and the strain energy

$$(2.11)\qquad \mathcal{E} = K + V = \frac{1}{2}\int_{-\ell/2}^{+\ell/2} \rho(x)A(x)\left[\frac{\partial w(x,t)}{\partial t}\right]^2 + E(x)I(x)\left[\frac{\partial^2 w}{\partial x^2}\right]^2 dx.$$

The physical interpretation of (2.9) and (2.10) implies that $\partial w/\partial t$ and $\partial^2 w/\partial x^2$ have to be square integrable on the interval $[-\ell/2,+\ell/2]$. We also assume that they are square integrable on $[0,T]$. We introduce the following product of two functions $u(x,t)$, $v(x,t)$, whose derivatives $\partial u/\partial t$, $\partial^2 u/\partial x^2$, $\partial v/\partial t$, $\partial^2 v/\partial x^2$ are square integrable functions in the interval $-\ell/2 \leq x \leq +\ell/2$:

$$(2.12)\qquad \langle u,v\rangle = \frac{1}{2}\int_{-\ell/2}^{+\ell/2}\left[\rho(x)\cdot A(x)\frac{\partial u}{\partial t}\frac{\partial v}{\partial t} + E(x)I(x)\frac{\partial^2 u}{\partial x^2}\frac{\partial^2 v}{\partial x^2}\right]dx.$$

$\langle u,v\rangle$ is clearly a function of t only. If $u = v$, then $\langle u,v\rangle$ is the total energy, as defined by the formulas (2.9), (2.10) and (2.11). A property of this product proved in Lemma 2.1 below will be of importance in the subsequent development of Pontryagin's principle.

<u>LEMMA 2.1.</u> <u>Let</u> u(x,t), v(x,t) <u>be</u> <u>two</u> <u>solutions</u> <u>of</u> <u>the</u> <u>MBVP</u> <u>with</u> <u>corresponding</u> <u>controls</u> f(x,t), g(x,t). <u>Then</u>

$$(2.13)\qquad \frac{d}{dt}\langle u,v\rangle = \frac{1}{2}\int_{-\ell/2}^{+\ell/2}\left(f(x,t)\frac{\partial v}{\partial t} + g(x,t)\frac{\partial u}{\partial t}\right)dx.$$

Before we prove this lemma, we emphasize that u(x,t) is the solution corresponding to the control f(x,t) and v(x,t) corresponds to g(x,t). f(x,t) and g(x,t) could be integrable and square integrable on $[-\ell/2,+\ell/2]$, or they could be Dirac delta functions. Despite the fact that the Dirac delta function is <u>not</u> a locally integrable function, we shall retain the commonly accepted use of the integral sign and interpret the resulting product as the usual linear map (see, for example, [15,pp.1-4]). No other changes will be necessary.

Proof of Lemma 2.1. We shall make use of the formulas (2.4a), (2.4b), and (2.4c) integrating by parts.

Thus

$$\frac{d}{dt} < u, v> = \frac{1}{2} \int_{-\ell/2}^{+\ell/2} \{ \rho(x) A(x) \left[\frac{\partial^2 u}{\partial t^2} \frac{\partial v}{\partial t} + \frac{\partial u}{\partial t} \frac{\partial^2 v}{\partial t^2} \right.$$

$$\left. + \frac{\partial}{\partial t} \left[E(x) I(x) \frac{\partial^2 u}{\partial x^2} \frac{\partial^2 v}{\partial x^2} \right] \right] \} \, dx$$

$$= \frac{1}{2} \int_{-\ell/2}^{+\ell/2} \{ \left[f(x,t) - \frac{\partial^2}{\partial x^2} \left(E(x) I(x) \frac{\partial^2 u}{\partial x^2} \right) \right] \cdot \frac{\partial v}{\partial t}$$

$$+ \left[g(x,t) - \frac{\partial^2}{\partial x^2} \left(E(x) I(x) \frac{\partial^2 v}{\partial x^2} \right) \right] \cdot \frac{\partial u}{\partial t}$$

$$+ \frac{\partial v}{\partial t} \cdot \left[\frac{\partial^2}{\partial x^2} \left(E(x) I(x) \frac{\partial^2 u}{\partial x^2} \right) \right]$$

$$+ v \cdot \frac{\partial}{\partial t} \left[\frac{\partial^2}{\partial x^2} \left(E(x) I(x) \frac{\partial^2 u}{\partial x} \right) \right] dx ,$$

$$= \frac{1}{2} \int_{-\ell/2}^{+\ell/2} \left[f(x,t) \frac{\partial v}{\partial t} + g(x,t) \frac{\partial u}{\partial t} \right] dx \qquad (2.13)$$

after integration by parts and interchange of the order of differentiation. (This formal manipulation is easily justified.)

<u>Corollary 1.</u> <u>If</u> $v(x,t)$ <u>is the solution of the homogeneous equation</u> (2.10), <u>then</u>

(2.14) $$\langle u,v\rangle_{t=\tau} = \langle u,v\rangle_{t=0} + \int_0^{\tau}\int_{-\ell/2}^{+\ell/2} f(x,t)\,\frac{\partial v}{\partial t}\,dx,\quad 0 \le \tau \le T.$$

Proof. Since v(x,t) is the solution of a homogeneous equation, $g(x,t) \equiv 0$, and

$$\langle u,v\rangle_{t=\tau} - \langle u,v\rangle_{t=0} = \int_0^{\tau}\int_{-\ell/2}^{+\ell/2} f(x,t)\,\frac{\partial v}{\partial t}\,dx$$

or

$$\langle u,v\rangle_{t=\tau} = \langle u,v\rangle_{t=0} + \int_0^{\tau}\int_{-\ell/2}^{+\ell/2} f(x,t)\,\frac{\partial v}{\partial t}\,dx$$

as required.

Corollary 2.2. If $f(x,t) \equiv g(x,t) \equiv 0$, then $\langle u,v\rangle \equiv$ constant. We observe that in the particular case when $u(x,t) = v(x,t)$ and $f(x,t) = g(x,t) \equiv 0$, this corollary reduces to the trivial statement that the total energy is conserved if the beam vibrates freely.

Lemma 2.2 (The Cauchy-Schwartz Inequality).

$$\langle u,v\rangle^2_{t=\text{const.}=\tau} \le \langle u,u\rangle\cdot\langle v,v\rangle\big|_{t=\tau},$$

and equality holds only if $\partial u/\partial t = c\,\partial v/\partial t, \partial^2u/\partial x^2 = c\,\partial^2 v/\partial x^2$ *for some constant* c.

Proof. It is sufficient to observe that <u,v> does satisfy all requirements of a scalar product for the "vectors" $[\partial u/\partial t, \partial^2 u/\partial x^2][\partial v/\partial t, \partial^2 v/\partial x^2]$, and that <u,u> is a norm (see, for example, [44,p.5]) for a classical proof).

2.4. Statement of the Control Problems.

Given the initial conditions (2.5a) and (2.5b) and one of the boundary conditions (2.4a), (2.4b), or (2.4c) at each boundary point $x = -\ell/2$, $x = +\ell/2$ and given $T > 0$, find an admissible control $\bar{\phi}(x,t)$ such that the total energy of the beam obeys the inequality

$$\mathcal{E}(\bar{\phi}(x,t), t=T) \leq \mathcal{E}(\phi(x,t), t=T), \tag{2.15}$$

where $\phi(x,t)$ is any other admissible control. The control $\bar{\phi}(x,t)$ will be called an optimal control for the interval [0,T].

The control problem stated above will be called the fixed interval control problem. Closely related to it is the minimal time control problem. Given the same initial and boundary conditions and given a nonnegative number E, such that $E < \mathcal{E}(t=0)$, find the control $\hat{\phi}(x,t)$ which reduces the total energy of the beam to the value E in the shortest possible time.

2.5. The Existence and Uniqueness Theorems for the fixed Interval Control Problem.

Theorem 2.1 (Existence of an optimal control). Let the MBVP as formulated by (2.1), (2.4), (2.5) be posed for an interval [0,T]. Then there exists at least one admissible optimal control $\tilde{\phi}(x,t)$ for the [0,T] fixed interval control problem.

Proof. Let E be the greatest lower bound on the total energy attainable at the time T through the use of admissible controls. (Clearly such a number exists.) Since $\mathcal{E}(T)$ depends continuously on the function $F(t) = \int_{-\ell/2}^{+\ell/2} \phi(x,t)dx$, where $\phi(x,t)$ is the control, we can choose a sequence of admissible controls $\phi_i(x,t)$ such that $\lim_{i\to\infty} \mathcal{E}(\phi_i(x,t),t=T) = E$. The generalized functions $\phi_i(x,t)$ obey the inequality

$$||\phi_i(x,t)||_{\Omega}^2 = \int_0^T (||\phi_i(x,t)||_x)^2 dt \leq 2T.$$

Hence the hypotheses of Appendix 1 are satisfied, and we can assert the existence of an admissible control $\phi(x,t) = \lim_{i\to\infty} \phi_i(x,t)$. It is easy to check that $\mathcal{E}(\phi(x,T),T) = E$, using (2.8) together with (2.9), (2.10), and (2.11).

Definition 2.1. The set of all functions $w(x,t)$, of the class C^1 in $\Omega = [-\ell/2,+\ell/2]\times[0,T]$, which are solutions of the MBVP, for which the inhomogeneous term is an admissible control, will be called an attainable set of displacements. The corresponding functions $E(x)I(x)(\partial^2 w(x,t)/\partial x^2)$ will be called the attainable bending moments and will be denoted by $M(x,t)$.

For convenience we introduce a new notation. We shall denote by $\hat{\tilde{W}}(x,t)$ the vector

$$\hat{\tilde{W}} = \left[\frac{\partial w(x,t)}{\partial t}, \frac{\partial^2 w(x,t)}{\partial x^2}\right]$$

We make the observation that the attainable set of displacements w forms a convex subset of $L_2\{[-\ell/2,+\ell/2]\times[0,T]$. This follows immediately from the linear dependence of displacements w(x,t) upon the controls (see (2.8)) and from the convexity of the admissible controls.

<u>Theorem 2.2</u> (Uniqueness of the finite state). <u>Let $\tilde{\phi}_1(x,t), \tilde{\phi}_2(x,t)$ be two admissible controls which are optimal controls for the [0,T] fixed interval. Then the corresponding displacement and velocity functions coincide at the time</u> $t = T$, i.e.,

$$\text{(2.16)} \quad \begin{aligned} w_1(\tilde{\phi}_1,x,T) &= w_2(\tilde{\phi}_2,x,T), \\ \frac{\partial w_1}{\partial t}(\tilde{\phi}_1,x,T) &= \frac{\partial w_2}{\partial t}(\tilde{\phi}_2,x,T). \end{aligned}$$

<u>Proof.</u> Let us assume to the contrary that there exist two optimal controls $\tilde{\phi}_1(x,t)$ and $\tilde{\phi}_2(x,t)$ such that

$$\hat{W}_1(\tilde{\phi}_1,x,T) \neq \hat{W}_2(\tilde{\phi}_2,x,T).$$

By the convexity of the attainable displacement set we conclude that $\frac{1}{2}(W_1 + W_2) = W(x,t)$ is also an attainable displacement. The corresponding total energy at the time T is

$$\mathcal{E}(\hat{W}(x,T)) = \frac{1}{2}\left\{\int_{-\ell/2}^{+\ell/2} \frac{\rho(x)A(x)}{4}\left[\frac{\partial(w_1+w_2)}{\partial t}\right]^2 + \frac{E(x)I(x)}{4}\left[\frac{\partial^2(w_1+w_2)}{\partial x^2}\right]^2 dx\right\}\Bigg|_{t=T},$$

$$= \frac{1}{4}\mathcal{E}(w_1(x,T)) + \frac{1}{4}\mathcal{E}(w_2(x,T)) + \frac{1}{2}\langle w_1,w_2\rangle_{t=T}$$

$$= \frac{1}{2}\tilde{E} + \frac{1}{2}\langle w_1,w_2\rangle_{t=T},$$

where $\tilde{E}$ is the minimal energy attainable at the time T by an application of an optimal control.

By Lemma 2.2 we have:

$$\langle w_1,w_2\rangle^2 \leq \langle w_1,w_1\rangle\cdot\langle w_2,w_2\rangle = \tilde{E}^2.$$

Since $\tilde{E}$ is the minimum total energy attainable, we must have

$\langle w_1,w_2\rangle^2 = \tilde{E}^2$. (Otherwise $\mathcal{E}(w,T) < \tilde{E}$, which is a contradiction.) But the strict equality

$$\langle w_1,w_2\rangle^2 = \langle w_1,w_1\rangle \cdot \langle w_2,w_2\rangle$$

implies that there exists a constant C such that

$$\frac{\partial w_1}{\partial t} = C\,\frac{\partial w_2}{\partial t}\,,\quad \frac{\partial^2 w_1}{\partial x^2} = C\,\frac{\partial^2 w_2}{\partial x^2}\,,\quad (\text{at } t = T).$$

Since $\mathcal{E}(\hat{w}_1,T) = \mathcal{E}(\hat{w}_2,T) = E$, upon substitution into (2.11) we find that $C = 1$, and that

$$\frac{\partial w_1}{\partial t} = \left.\frac{\partial w_2}{\partial t}\right|_{t=T} \quad \text{and} \quad \left.\frac{\partial^2 w_1}{\partial x^2}\right|_{t=T} = \left.\frac{\partial^2 w_2}{\partial x^2}\right|_{t=T}$$

for almost all $x \in [-\ell/2,+\ell/2]$.

From the assumption of piecewise smoothness of $\partial^2 w/\partial x^2$, we conclude that

$$w_2(x,T) = w_1(x,T) + C_1(x) + C_2,$$

where C_1 and C_2 are constants. (An identical conclusion would follow a more general hypothesis that $\partial^2 w/\partial x^2$ could be a step function with a finite number of steps in $[-\ell/2,+\ell/2]$, and the formula is clearly valid when $\partial^2 w/\partial x^2$ is a sum of such a step function and of a piecewise smooth function. This problem does not arise with $\partial w/\partial t$ which must be continuous in $[-\ell/2,+\ell/2]$.)

Assuming that either conditions (2.4b) or (2.4c) are applicable at one end of the beam, and either (4a), (4b), or (4c) at the other end, we see that $C_1 = C_2 = 0$. In the remaining case, when conditions (4a) are applied at both ends of the beam, we also arrive at the conclusion that $C_1 = C_2 = 0$. Then in all possible cases $C_1 = C_2 = 0$, and $\underline{w}_1(x,T) \equiv \underline{w}_2(x,T)$. This completes the proof.

2.6 Pontryagin's principle. The principle stated in Theorem 3 below is in complete agreement with the maximum principle of Pontryagin (see [28]), and also with the results of [32] and [17]. The proof parallels the proof of Russell [32] for the symmetric hyperbolic systems with a few differences.

Theorem 3. *Let $\tilde{\phi}(x,t)$ be an optimal control for the fixed interval control of the MBVP, as stated in the preceding section. We assume that $\mathcal{A}(\tilde{\phi}(x,t),t) > 0$ if $t \in [0,T]$. Let $\tilde{w}(x,t)$ be the solution of the MBVP corresponding to $\tilde{\phi}(x,t)$. Let $v(x,t)$ be a solution of the homogeneous equation (2.1a) satisfying the same boundary conditions, and such that $v(x,T) = \tilde{w}(x,T)$, i.e.*

$$v(x,T) = w(x,T),$$

$$\frac{\partial v(x,T)}{\partial t} = \frac{\partial w(x,T)}{\partial t}.$$

<u>Then</u>

(2.17) $$\int_{-\ell/2}^{+\ell/2} [-\tilde{\phi}(x,t)\frac{\partial v(x,t)}{\partial t}]dx = \max \int_{-\ell/2}^{+\ell/2} [-\phi(x,t)\frac{\partial v(x,t)}{\partial t}]dx$$

<u>for all admissible controls</u> $\phi(x,t)$.

<u>Proof.</u> If we can prove the theorem under the additional assumption that all controls including $\tilde{\phi}(x,t)$ are piecewise continuous (piecewise smooth), the general statement will follow as an easy corollary. Let $t = \tau$ be a point of continuity of the optimal control $\tilde{\phi}(x,t)$, $\tau \in (0,T)$. Then there exists $\sigma > 0$, such that $\tilde{\phi}(x,t)$ is continuous in the interval $I_\sigma = [\tau - \sigma, \tau + \sigma]$, and the interval I_σ is contained in $[0,T]$. Let $\psi(x,t)$ be an admissible control such that for a sufficiently small number $\varepsilon > 0$, $\tilde{\phi} + \varepsilon\psi$ is an admissible control. (Clearly, if no such $\psi(x,t)$ can be found, $\tilde{\phi}(x,t)$ is the unique admissible control in I_σ, and there is nothing to prove.)

Let us consider the control

$$\phi'(x,t) = \begin{cases} \tilde{\phi}(x,t) & \text{for } t \in [0,T] - I_\sigma, \\ \tilde{\phi}(x,t) + \varepsilon\psi(x,t) & \text{for } t \in I_\sigma. \end{cases}$$

$\phi'(x,t)$ is clearly an admissible control.

Let ψ_σ be the control:

$$\psi_\sigma(x,t) = \begin{cases} 0 & \text{for } t \notin I_\sigma, \\ \psi(x,t) & \text{for } t \in I_\sigma. \end{cases}$$

σ can be chosen so that ψ_σ is smooth in I_σ. (All controls are piecewise smooth functions of time.)

Let $w(\phi,x,t)$ be the solution of the MBVP corresponding to the control $\tilde{\phi}(x,t)$, and $\tilde{w}_\sigma(\psi_\sigma,x,t)$ be the solution of the MBVP corresponding to the control ψ_σ with the same boundary conditions but with <u>zero initial conditions</u>. (We do not wish to add w_H to both terms of the right-hand side in (2.18) below.)

The solution $w'(x,t)$ corresponding to the control $\phi'(x,t)$ is

$$\text{(2.18)} \qquad w'(x,t) = \tilde{w}(x,t) + \varepsilon w_\sigma(x,t).$$

The expression for total energy is

$$\text{(2.19)} \qquad \mathcal{E}(w'(x,t),t) = \mathcal{E}(\tilde{w}(x,t),t) + 2\varepsilon\langle w,w_0\rangle + \varepsilon^2\mathcal{E}(w_\sigma(x,t),t).$$

Let w_σ be such that w' is not optimal displacement:

$$\mathcal{E}(w'(x,t),T) > \mathcal{E}(\tilde{w}(x,t),T). \tag{2.20}$$

Then $\langle w,w_\sigma\rangle_{t=T} \geq 0$.

Since ε was arbitrary, and the total energy is a continuous function of time, there must also exist an interval $[T-\varepsilon,T]$ such that

$$\langle \tilde{w},w_\sigma\rangle \geq 0 \quad \text{for all } t \in [T-\varepsilon,T].$$

By Corollary 2.2 of Lemma 2.1, $\langle v,w_\sigma\rangle$ = const. on the interval $[t+\sigma,T]$ since in this interval $w_\sigma(x,t)$ is a solution of the homogeneous MBVP, while $v(x,t)$ is a solution of the homogeneous MBVP by hypothesis.

Hence,

$$\langle v,w_\sigma\rangle_{t=\tau+\sigma} = \langle v,w_\sigma\rangle_{t=T} = \langle w,w_\sigma\rangle_{t=T} \geq 0.$$

In the interval $[0,\tau-\sigma]$, we have $\langle \tilde{w},w_\sigma\rangle \equiv 0$ since $w_\sigma \equiv 0$ in that interval. In the interval I_σ we consider the limit:

$$\lim_{\sigma\to 0} \frac{1}{\sigma}\int_{\tau-\sigma}^{\tau+\sigma} \langle v,w_\sigma\rangle dt = \lim \frac{1}{\sigma}\int_{\tau-\sigma}^{\tau+\sigma}\int_{-\ell/2}^{+\ell/2} [\psi_\sigma \frac{\partial v(x,t)}{\partial t}]dxdt = 0$$

uniformly, since τ was a point of continuity of ψ_σ, and $\partial v/\partial t$,

ψ_σ are smooth functions of time in I_σ. Consequently, $\langle v,w_\sigma \rangle \geq 0$ in I_σ. Collecting these results, we have $\langle v,w_\sigma \rangle \geq 0$ in [0,T].

Using (2.13) and (2.14) we have

$$\int_{-\ell/2}^{+\ell/2} [-\tilde{\phi}(x,t) \frac{\partial v(x,t)}{\partial t}]dx = \max \int_{-\ell/2}^{+\ell/2} [-\phi(x,t) \frac{\partial v(x,t)}{\partial t}]dx$$

for all admissible controls $\phi(x,t)$ and for all $t \in [0,T]$. The proof is complete since τ, σ were arbitrary and ϕ', $\tilde{\phi}$ were piecewise smooth functions of time, and any admissible control could be obtained by altering $\tilde{\phi}(x,t)$ on a collection of suitable intervals I_σ.

Note. The proof is unchanged if $t = \tau$ is chosen to be a regular point in the sense of Pontryagin, rather than a point of continuity in [0,T]. Hence the theorem goes through with weaker hypothesis regarding $\tilde{\phi}(x,t)$, but with more complex arguments.

2.7. Application of Theorem 2.3. At first it seems that the result is only of theoretical interest, since in an attempt at direct application we need to know the finite state of the beam following an optimum control to decide if the control was optimal. This situation arises in a number of physical problems and is usually dealt with by some iterative schemes. The immediate value of the maximum principle as stated in

Theorem 2.3 is in providing negative answers to the question: Is a proposed control optimal? An easy example of such application is given below.

Consider a beam which is simply supported at both ends, that is, conditions (2.4b) are satisfied at the points $x = \pm\ell/2$. The initial condition is given by

$$w(x,0) = \frac{p}{24E\cdot I}\left[\ell^3\left(x + \frac{\ell}{2}\right) - 2\ell\left(x + \frac{\ell}{2}\right)^2 + \left(x + \frac{\ell}{2}\right)^4\right],$$

where p, E and I are constant, $\partial w(x,0)/\partial t \equiv 0$. This represents the case of an initial deflection due to a constant load p (say due to a wind load), with the load being suddenly removed at the time $t = 0$.

The fundamental frequency of the beam is given by

$$n_1 = \frac{\pi}{2} \frac{\sqrt{g}\, EI}{p\ell^4} .$$

We now propose what would intuitively appear to be a "good" control. Among the piecewise continuous functions $\phi(x,t)$ which have the property $\int_{-\ell/2}^{+\ell/2} \phi(x,t)dx \leq p\cdot\ell$ for all $t \geq 0$, we select the control

$$\phi(x,t) = -p\cdot\mathrm{sgn}\left(\frac{\partial w_H}{\partial t}\right) ,$$

where the sign function is given by

$$\operatorname{sgn} y = \begin{cases} -1 & \text{if } y < 0, \\ 0 & \text{if } y = 0, \\ +1 & \text{if } y > 0, \end{cases}$$

while $w_H(x,t)$ is given by the formula:

$$w_H = w(x,0)\cos 2\pi n_1 t,$$

and hope that this control is optimal for the interval $[0,n_1/2]$.

This is seen to be incorrect (without even applying Pontryagin's principle), since this control amounts to an immediate restoration of the static load, and the total energy of the beam will remain constant. Clearly a better control is attained by assuming

$$\begin{cases} \phi(x,t) \equiv 0 & 0 \le t \le \frac{n_1}{4}, \\ \phi(x,t) = -p \operatorname{sgn}\left(\frac{\partial w_H}{\partial t}\right), & \frac{n_1}{4} < t \le \frac{n_1}{2}. \end{cases}$$

The total energy dissipated by application of this control is given by

$$\mathcal{E}_D = \int_{n_1/4}^{n_1/2} \int_{-\ell/2}^{+\ell/2} \phi(x,t)\frac{\partial w_H(x,t)}{\partial t}\,dxdt$$

The initial energy of the beam is given by

$$\mathcal{E}(0) = \frac{1}{2}\int_{-\ell/2}^{+\ell/2} EI\left(\frac{\partial^2 w(x,0)}{\partial x^2}\right)^2 dx$$

and the final energy by

$$\mathcal{E}\left(\frac{n_1}{2}\right) = \frac{1}{2}\int_{-\ell/2}^{+\ell/2} EI\left(\frac{\partial^2 w(x,n_1/2)}{\partial x^2}\right) dx.$$

Then

$$w(x,\frac{n_1}{2}) = \frac{p_1}{24EI}\left(\ell^3(x+\frac{\ell}{2}) - 2\ell(x+\frac{\ell}{2})^2 + (x+\frac{\ell}{2})^4\right),$$

where p_1 is computed from the relationship

$$\mathcal{E}(n_1/2) = \mathcal{E}(0) - \mathcal{E}_D.$$

The functions $w_H(x,t)$ and $v(x,t)$ of Theorem 2.3 are scalar multiples of each other, so that in the interval $[n_1/4, n_1/2]$, $\phi(x,t)$ may well be optimal. However, it is not optimal in the interval $[0, n_1/4]$ where any force $f(x,t)$ equal to $-(p\cdot\ell\cdot\mathrm{sgn}(\partial v/\partial t))$ on a subinterval $[\tau-\sigma, \tau+\sigma]$ with σ sufficiently small, and equal to zero otherwise, results in

$$\int_{-\ell/2}^{+\ell/2} -\left(f(x,t)\frac{\partial v}{\partial t}\right) > 0.$$

Hence, $\phi(x,t) \equiv 0$ is not optimal on the subinterval $[0,n_1/4]$. We can now easily modify $\phi(x,t)$ to improve the control on the subinterval $[n_1/8,n_1/4]$, etc. It is clear how the maximal principle can be used to effect a gradual improvement of some arbitrarily chosen control. A more systematic approach using nonlinear programming techniques has been discussed in [32]. The inconvenience of having to compute the final state before being able to check the optimality is obvious.

In the next section of this paper we shall introduce a criterion of instant optimality, which results as before in a form of Pontryagin's principle requiring the knowledge of only the initial and currect displacements.

2.8. Some general discussion. We stress the assumption $\mathcal{S}(\tilde{\phi}(x,t),t) > 0$ for all $t \in [0,T]$ of Theorem 2.3. If $\mathcal{S}(\dot{\phi}(x,T),T) = 0$, Theorem 3 is meaningless, because $v(x,t) \equiv 0$. We shall be able to restate the maximal principle in a meaningful way for the case when $\mathcal{S}(\dot{\phi}(x,t),t) > 0$ when $t \in [0,T)$, allowing $\mathcal{S}(\tilde{\phi}(x,t),T)$ to be equal to zero. However, if $\mathcal{S}(\tilde{\phi}(x,t),t) \equiv$ const. on some subinterval of $[0,T]$, it appears to be impossible to patch up the difficulties.

In particular, we want to avoid the situation demonstrated by Fig. 1, Chapter 1. Consider a beam vibrating freely, so that the homogeneous equation (2.1) has a solution $\underset{\sim}{w}_H(x,t)$ such that $\underset{\sim}{w}_H(x,0) = \underset{\sim}{w}_H(x,t_1)$. If we choose the value T large enough, it is possible to apply two optimal controls for the interval $[0,T]:\phi_1(x,t)$ and $\phi_2(x,t)$, such that $\phi_2(x,t+t_1) = \phi_1(x,t)$ when $t > t_1$, $\phi_2(x,t) \equiv 0$ when $t < t_1$, and $\mathcal{E}(\phi_1,T) = \mathcal{E}(\phi_2,T) = 0$. We specially want to avoid optimal controls like $\phi_2(x,t)$.

2.9. Instantly optimal controls of vibrating beams. The definition of an instantly optimal control is identical with the one given in Chapter 1 for the symmetric hyperbolic systems. As before the instantly optimal control $\hat{\phi}(x,t)$ is unique.

Theorem 2.4. The instantly optimal control $\hat{\phi}(x,t)$ satisfies the maximum principle:

$$\int_{-\ell/2}^{+\ell/2} [-\hat{\phi}(x,t) \frac{\partial w(\hat{\phi},x,t)}{\partial t}]dx = \max \int_{-\ell/2}^{+\ell/2} [-\phi(x,t) \frac{\partial w(\phi,x,t)}{\partial t}]dx$$

for all admissible controls $\phi(x,t)$, and for any t in the domain of the control $\varphi(x,t)$, such that $E(t) > E$.

Proof: The proof parallels the proof given in Chapter 1 for the symmetric hyperbolic case.

It is easy to give examples of instantly optimal controls which are not optimal over a fixed time interval. Consider the initial conditions

$$w(x,0) \equiv 0,\ x \in [-\ell/2, +\ell/2],$$

$$\frac{\partial w(x,0)}{\partial t} = \frac{p}{24EI}\left[\ell^3\left(x+\frac{\ell}{2}\right) - 2\ell^2\left(x+\frac{\ell}{2}\right)^2 + \left(x+\frac{\ell}{2}\right)^4\right]$$

(Corresponding to a uniform load case). The time optimal control $\hat{\phi}(x,t)$ will satisfy the relationship (2.17), i.e.,

$$\int_{-\ell/2}^{+\ell/2} -\hat{\phi}(x,t)\,\frac{\partial v(x,t)}{\partial t}\,dx = \max \int_{-\ell/2}^{+\ell/2} -\phi(x,t)\,\frac{\partial v(x,t)}{\partial t}\,dx,$$

$$||\phi|| \leq K.$$

On the other hand, the instantly optimal control would satisfy

$$\int_{-\ell/2}^{+\ell/2} -\phi(x,t)\,\frac{\partial w(\hat{\phi},x,t)}{\partial t}\,dx = \max \int_{-\ell/2}^{+\ell/2} -\phi(x,t)\,\frac{\partial w(\phi,x,t)}{\partial t}\,.$$

It is easy to show that the instantly optimal control at the time $t = 0$ is the Dirac delta function (a point load) of a given norm K (which is chosen a priori) applied at the point $x = 0$.).

Consider some time $t = \tau$ during which the beam has been deformed by the application of point loads applied to the points of highest velocity of the beam. Letting the beam vibrate freely back to the initial time $t = 0$, we can check numerically that in general a completely different configuration of the beam $w(x,0) \neq 0$ is obtained, and in general the product

$$\int_{-\ell/2}^{+\ell/2} \hat{\phi} \cdot \frac{\partial v(x,t)}{\partial t} \, dx \Big|_{(t=0)}$$

will not coincide with

$$\int_{-\ell/2}^{+\ell/2} \hat{\phi} \frac{\partial w(\hat{\phi},x,t)}{\partial t} \, dx, \Big|_{(t=0)}$$

showing that $\hat{\phi}(x,t)$ was not time optimal for the interval $[0,\tau]$.

2.10. Optimal excitation of beams. A class of problems closely related to the optimal control problems shall be referred here as the optimal excitation. The definition given below (definition 1) is of the optimal excitation over fixed time interval. In analogy with the optimal control problems we can define time optimal excitation, and instantly optimal excitation.

Definition 1

Given the initial and boundary conditions, let $\tilde{\phi}(x,t)$ be an admissible control on the interval $[0,T]$, such that $\mathscr{E}(\tilde{\phi}(x,t);T) \geq \mathscr{E}(f(x,t);T)$ for any admissible control f. Then $\tilde{\phi}(x,t)$ will be called the optimal excitation control (or optimal excitation) for the fixed time interval $[0,T]$.

Definition 2

An admissible control $\tilde{\phi}(x,t)$ such that the total energy $\mathcal{E}(t)$ will reach a fixed value $\mathcal{E} > \mathcal{E}(0)$ in the shortest possible time will be called time optimal excitation.

2.11. The Pontryagin's principle. The following propositions are easily proved following arguments similar to those given for the optimal control case.

Proposition A

There exists at least one time optimal excitation control, and at least one optimal excitation for a fixed time interval $[0,T]$. (Since all controls are square integrable functions of time, the existence of such control follows easily from the completeness of the L_2 spaces.)

Corollary A

Given any initial conditions at $t = t_1$, and given any value $t_2 > t_1$ there exists some control $\phi(x,t)$ such that $\mathcal{E}(\phi(x,t);t_2) > \mathcal{E}(\phi(x,t);t_1)$.

Proposition B

Every optimal excitation for a fixed time interval $[0,T]$ is also time optimal.

Proof

Assume to the contrary that $\tilde{\phi}(x,t)$ is optimal for the fixed time interval $[0,T]$, and increases the total energy to the maximum level $\tilde{\mathcal{E}}(T)$, but that $\tilde{\phi}$ is not time optimal. Hence there would exist another admissible control $\bar{\phi}(x,t)$ which would increase the energy level to $\tilde{\mathcal{E}}$ at the time $\hat{\tau} < T$. The use of corollary A (using $T = t_2$, $\hat{\tau} = t_1$) now leads to a contradiction.

Theorem 1

Let either of the following boundary conditions be obeyed at either end: $x = \pm\frac{\ell}{2}$: a) free support, b) built in, c) free end, and let the initial conditions $w(x,0) = w_0(x)$, $\frac{\partial w(x,0)}{\partial t} = v_0(x)$ be given. Let $\tilde{\phi}(x,t)$ be an optimal excitation for the fixed time interval $[0,T]$. Then

$$(2.21) \qquad \int_{-\ell/2}^{+\ell/2} f(x,t) \frac{\partial w_H(x,t)}{\partial t} dx \le \int \tilde{\phi}(x,t) \frac{\partial w_H(x,t)}{\partial t} dx$$

for any admissible control $f(x,t)$. $w_H(x,t)$ denotes a solution of the homogeneous equation (2.1a) satisfying the same final conditions as $\tilde{w}(\tilde{\phi}(x,t);x,t)$, i.e.:

$$w_H(x,T) = \tilde{w}(\tilde{\phi}(x,t);x,T)$$

$$\frac{\partial w_H(x,T)}{\partial t} = \frac{\partial \tilde{w}(\tilde{\phi}(x,t);x,T)}{\partial t}$$

The proof parallels the one given for the optimal control with all inequalities reversed.

The proof is routine, once the argument has been seen in [28] or in [32]. There is a basic difference however in the result given by [17] and the one given by our formula (2.21) above, despite the fact that the two formulae appear to be identical except for the reversal of the inequality sign. We lack here the analogue of the Theorem 2 of [17], or the Theorem 2.2 of this work. (Theorem 2.2 has stated that if $\tilde{\phi}_1, \tilde{\phi}_2$ are two optimal controls for a fixed time interval $[0,T]$, then

$$w_1(\tilde{\phi}_1(x,t);x,T) = w_2(\tilde{\phi}_2(x,t);x,T)$$

and

$$\frac{\partial w_1(x,T)}{\partial t} = \frac{\partial w_2(x,T)}{\partial t},$$

(that is even if the optimal controls are not unique, the resulting final state is unique.) To prove this we had to use the convexity of optimal controls: if $\tilde{\phi}_1(x,t)$ and $\tilde{\phi}_2(x,t)$ are optimal, then $\Lambda\tilde{\phi}_1(x,t) + (1-\Lambda)\tilde{\phi}_2(x,t)$ is also optimal for any $0 \leq \Lambda \leq 1$. For the optimal excitation such lemma has not been given. We can prove, however, a contrapositive statement concerning convexity, which we shall label:

Proposition C

Let $\phi_1(x,t)$, $\phi_2(x,t)$ be two optimal excitations, such that either $w_1(\phi_1(x,t);T) \neq w_2(\phi_2(x,t);T)$ or $\frac{\partial w_1(\phi_1(x,t);T)}{\partial t} \neq \frac{\partial w_2(\phi_2(x,t);T)}{\partial t}$. Then $\tilde{\phi}(x,t) = \Lambda\phi_1 + (1-\Lambda)\phi_2$, $0 < \Lambda < 1$, can not be an optimal excitation.

Proof

Consider $\mathcal{E}(\phi(x,t);t) = \Lambda^2\mathcal{E}(\phi_1(x,t),t) + (1-\Lambda)^2\mathcal{E}(\phi_2(x,t);t) + 2\Lambda(1-\Lambda) \langle w_1(\phi_1),w_2(\phi_2)\rangle_t$. In particular at the time T we have $\mathcal{E}(\phi_1(x,t),T) = \mathcal{E}(\phi_2(x,t),T) = \max (\mathcal{E}(f);T) = \tilde{\mathcal{E}}$, for all admissible controls f. Hence:

$$\mathcal{E}(\phi(x,t);T) = \Lambda^2\tilde{\mathcal{E}} + (1-\Lambda)^2\tilde{\mathcal{E}} + 2\Lambda(1-\Lambda)\langle w_1,w_2\rangle_{t=T} ,$$

$$\begin{aligned}\mathcal{E}(\phi(x,t);T) &= -2\Lambda(1-\Lambda)\tilde{\mathcal{E}} + \tilde{\mathcal{E}} + 2\Lambda(1-\Lambda)[\langle w_1,w_2\rangle_{t=T}].\\ &= \tilde{\mathcal{E}} - 2\Lambda(1-\Lambda)(\mathcal{E} - \langle w_1,w_2\rangle_{t=T}).\end{aligned}$$

Since $0 < \Lambda < 1$, the coefficient $2\Lambda(1-\Lambda)$ is positive.

We now use the easily proved inequality: (the Cauchy-Schwartz inequality for the scalar product $\langle w_1,w_2\rangle$)

$\langle w_1,w_2\rangle_T^2 \leq \langle w_1,w_1\rangle_T \cdot \langle w_2,w_2\rangle_T = \tilde{\mathcal{E}}^2$, from which follows the result: $\mathcal{E}(\phi(x,t);T) \leq \bar{\mathcal{E}}$. The equality is valid only if $w_1(x,T) = w_2(x,T)$ and $\frac{\partial w_1(x,T)}{\partial t} = \frac{\partial w_2(x,T)}{\partial t}$ - a condition which has been denied by our hypothesis. Hence, we have $\mathcal{E}(\phi(x,t);T) < \tilde{\mathcal{E}}$, and $\phi(x,t)$ is not optimal, which was to be proved.

Proposition D

Let ϕ_1, ϕ_2 be two optimal excitations on a fixed time interval $[0,T]$. Let w_{Hi} denote a solution of the homogeneous equation (2.1a), such that

$$w_{Hi}(x,T) = \tilde{w}(\phi_i(x,t);T)$$

$$\frac{\partial w_{Hi}(x,T)}{\partial t} = \frac{\partial \dot{w}(\phi_i(x,t);T)}{\partial t}, \quad i = 1,2.$$

Then

$$\int_{-\ell/2}^{+\ell/2} [(\phi_1 - \phi_2) \frac{\partial (w_{H1} - w_{H2})}{\partial t}]dx \geq 0.$$

Proof

If ϕ_1, ϕ_2 are two optimal excitations for the fixed interval $[0,T]$, then according to the formula (2.21) at any time $t \varepsilon [0,T]$ we must have:

$$\int_{-\ell/2}^{+\ell/2} \phi_1(x,t) \frac{\partial_{H1}(x,t)}{\partial t} dx \geq \int_{-\ell/2}^{+\ell/2} \phi_2(x,t) \frac{\partial_{H1}(x,t)}{\partial t} dx$$

and

$$\int_{-\ell/2}^{+\ell/2} -\phi_1(x,t)\frac{\partial w_{H2}(x,t)}{\partial t}\,dx \geq \int_{-\ell/2}^{+\ell/2} -\phi_2(x,t)\frac{\partial w_{H2}(x,t)}{\partial t}\,dx.$$

Adding left and right hand sides of these inequalities we have:

$$\int_{-\ell/2}^{+\ell/2} \phi_1\left[\frac{\partial w_{H1}}{\partial t} - \frac{\partial w_{H2}}{\partial t}\right]dx \geq \int_{-\ell/2}^{+\ell/2} \phi_2\left[\frac{\partial w_{H1}}{\partial t} - \frac{\partial w_{H2}}{\partial t}\right]dx.$$

This is equivalent to:

$$\int_{-\ell/2}^{+\ell/2} (\phi_1 - \phi_2)\,\frac{\partial (w_{H1}-w_{H2})}{\partial t}\,dx \geq 0,$$

which is the required inequality.

We comment that there is no reason to expect uniqueness of either optimal excitation or of the final state of the beam. We would also like to comment that the arguments given above have used very little information pertaining to beam theory, and are quite easily translated to hyperbolic systems, to plate theory, or other related problems.

2.11 The optimal control of coupled transverse and torsional vibrations of beams.

The previous discussion deriving the basic maximality principles for vibrating beams is inapplicable in the case of coupled bending and torsional vibrations, which case is of particular importance in the dynamics of aircraft structures. This section proves the existence, uniqueness and the maximality principle for optimal controls of the coupled bending and torsional vibrations of beams. The warping effects have been taken into account in this theory.

2.11.1 Notation and basic definitions

We shall consider the vibration of a beam governed by the pair of differential equations:

$$(2.22\underline{a})\quad L_1(y,\theta) = \frac{\partial^2}{\partial x^2}\left[E(x)I(x)\frac{\partial^2 y(x,t)}{\partial x^2}\right] + \rho(x)A(x)\frac{\partial^2 y(x,t)}{\partial t^2} - \rho(x)A(x)e(x)\frac{\partial^2 \theta(x,t)}{\partial t^2} = \phi_1(x)u_1(t)$$

$$(2.22\underline{b})\quad L_2(y,\theta) = \frac{\partial^2}{\partial x^2}\left[E(x)C_w(x)\frac{\partial^2 (x,t)}{\partial x^2}\right] - \frac{\partial^2}{\partial x^2}\left[G(x)C(x)\,\theta(x,t)\right] - \rho(x)A(x)e(x)\frac{\partial^2 y(x,t)}{\partial t^2} - \rho(x)I_0(x)\frac{\partial^2 (x,t)}{\partial t^2} = \phi_2(x)u_2(t).$$

The symbols used above have the following physical meaning:

$y(x,t)$ - the transverse displacement;

$\theta(x,t)$ - the angle of rotation of the cross-section;

$e(x)$ - the distance from the centroid to the center of torsion;

$y_g(x,t)$ ($y_g = y - e\theta$) — the transverse displacement of the centroid of the cross-section;

$A(x)$ - the cross-sectional area;

$I_p(x)$ - the polar moment of inertia of the cross-section about the centroid;

$I_0(x)$ ($I_0 = I_p + Ae^2$) — the polar moment of inertia with respect to the shear center.

$E(x)$ - Young's modulus;

$\rho(x)$ - the material density;

C_w - the modulus of warping rigidity;

(EC_w is the warping rigidity of the beam)

G - modulus of torsional rigidity;

GC - denotes the torsional rigidity, and

C is the torsional constant, satisfying $T_t = GC \frac{\partial\theta}{\partial x}$

where T_t is the twisting moment due to shear stresses;

$\phi_1(x)u_1(t)$ - are the forces applied to the beam;

$\phi_2(x)u_2(t)$ the first has the physical dimension of force per unit length, the second of moment per unit length.

We assume that the applied force (i.e. the inhomogeneous

term) is of the form $\phi_i(x)u_i(t)$, $i = 1,2$, where $\phi_i(x)$ is either a bounded and square integrable, (hence absolutely integrable function of x) on the interval $[0,\ell]$, or is either the Dirac delta function or its derivative concentrated on a finite number of points of $[0,\ell]$. That is,

$$\phi_j(x) = \sum_{i=1}^{K} \alpha_{ij}\delta(x - \xi_i) + \sigma_j(x),\ j = 1,2 \quad \text{where}$$

α_{ij} are constants, $\sigma_j(x)$ is a uniformly bounded, square integrable function on $[0,\ell]$.

There may be a restriction of the type:

$$\sum_{i=1}^{K} |\alpha_{ij}| + \int_0^{\ell} |\sigma_j(x)|dx \leq C \text{ for some specified}$$

constant C, in the a priori choice of α_{ij}, ξ_i, $\sigma_j(x)$. $u_i(t)$ is a piecewise continuous function on $[0,T]$, satisfying $||u_i|| \leq 1$. u_i, $i = 1,2$, are to be determined to optimize the energy level as postulated in the control problem.

We shall consider here only the problem of constant E, ρ, I, A, e, hence of constant I_0, C_w on $[0,\ell]$. This is done for reasons of simplicity. The basic discussion can be easily generalized to the case where EI, ρA, EC_w, GC, ρI_0, e are arbitrary piecewise continuous functions of x. The modified equations $(2.22\tilde{\underline{a}})$, $(2.22\tilde{\underline{b}})$ are:

(2.22a) $$L_1(y,\theta) = EI\,\frac{\partial^4 y(x,t)}{\partial x^4} + \rho A\,\frac{\partial^2 y(x,t)}{\partial t^2}$$

$$- \rho Ae\,\frac{\partial^2 \theta(x,t)}{\partial t^2} = \phi_1(x)u_1(t).$$

(2.22b) $$L_2(y,\theta) = EC_w\,\frac{\partial^4 \theta(x,t)}{\partial x^4} - GC\,\frac{\partial^2 \theta(x,t)}{\partial x^2} - \rho Ae$$

$$\frac{\partial^2 y(x,t)}{\partial t^2} - \rho I_0\,\frac{\partial^2 \theta(x,t)}{\partial t^2} = \phi_2(x)u_2(t).$$

The beam has an axis of symmetry, here the x-axis, with the y-axis passing through the shear center, and the x-axis coinciding with the centroidal axis, as illustrated on figure 2.1.

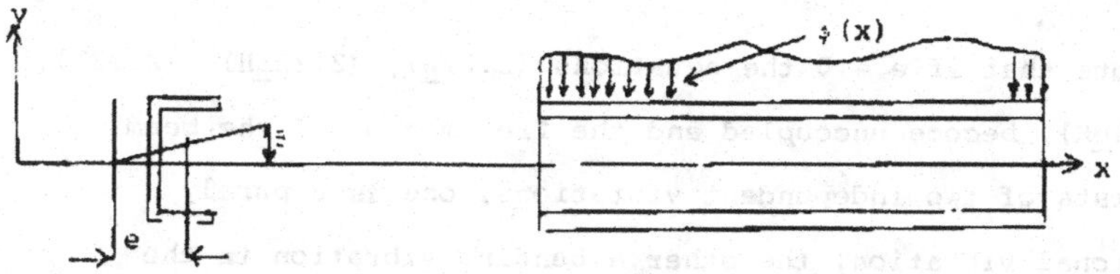

Figure 2.1.

If a load w(x) is distributed along the x-axis of the beam, it is causing a torque of magnitude $w(x) \cdot e$ to be distributed along the beam. Let us now assume that the only load applied is the inertia of the beam, and the beam is vibrating freely. Linearalized equations of a free motion are:

(2.22aH) $\quad L_1(y,\theta)$

$$= EI \frac{\partial^4 y}{\partial x^4} + \rho A \frac{\partial^2 y}{\partial t^2} - \rho Ae \frac{\partial^2 \theta}{\partial t^2} = 0 .$$

(2.22bH) $\quad L_2(y,\theta)$

$$= EC_w \frac{\partial^4 \theta}{\partial x^4} - GC \frac{\partial^2 \theta}{\partial x^2} - \rho Ae \frac{\partial^2 y}{\partial t^2} - \rho I_0 \frac{\partial^2 \theta}{\partial t^2}$$

$$= 0.$$

We note that if e = 0 the equations (2.22a), (2.22aH), (2.22b), (2,22bH), become uncoupled and the free motion of the beam consists of two independent vibrations, one is a purely torsional vibration, the other a bending vibration in the y-direction. In general the motion will be a coupled vibration with energy transfer taking place between the two modes of vibrations.

The total potential energy is:

(2.23a) $$V = \frac{EI}{2}\int_0^\ell \left(\frac{\partial^2 y}{\partial x^2}\right)^2 dx + \frac{GC}{2}\int_0^\ell \left(\frac{\partial\theta}{\partial x}\right)^2 dx + \frac{EC_w}{2}\int_0^\ell \left(\frac{\partial^2\theta}{\partial x^2}\right)^2 dx,$$

and the kinetic energy is:

(2.23b) $$T = \rho\frac{A}{2}\int_0^\ell \left(\frac{\partial y_g}{\partial t}\right)^2 dx + \rho\frac{I_p}{2}\int_0^\ell \left(\frac{\partial\theta}{\partial t}\right)^2 dx$$

where y_g is the y coordinate of the centroid of cross-section of the beam. If we assume the approximate relationship $y_g = y - e\theta$, we obtain

$$\frac{\partial y_g}{\partial t} = \frac{\partial y}{\partial t} - e\frac{\partial\theta}{\partial t} \quad ,$$

which can be substituted into the equation (2.23b).

The total energy is defined by the equation:

(2.24) $$\mathcal{E}(t) = T + V \quad .$$

2.11.2. The Boundary conditions

We note that warping and longitudinal stresses in the beam are proportional to

$$\frac{\partial\theta}{\partial x} \quad , \quad \frac{\partial^2\theta}{\partial x^2}$$

respectively. As an example we could consider a simple support (at $x = 0$, or $x = \ell$), which implies restraint against rotation; however, the beam is free to warp at the support point. Hence, we have at:

$$\begin{cases} x = 0 \\ \text{or} \\ x = \ell \end{cases} \qquad \begin{cases} \theta = 0 \\ \dfrac{\partial^2 \theta}{\partial x^2} = 0 \end{cases}$$

If the support is of the built-in type, no warping deformation is possible and we have:

$$\begin{cases} \theta = 0 \\ \dfrac{\partial \theta}{\partial x} = 0 \end{cases} .$$

Combining these with the usual support conditions for $w(x,t)$, we now postulate the boundary conditions of either (2.25a) or (2.25b) type at the boundary points $x = 0$ and $x = \ell$. Let b denote either 0, or ℓ, then we have either:

<u>Simple support</u>

(2.25a)

$$\left.\begin{aligned} y(b,t) &\equiv 0 \\ \frac{\partial^2 y(b,t)}{\partial x^2} &\equiv 0 \\ \theta(b,t) &\equiv 0 \\ \frac{\partial^2 \theta(b,t)}{\partial x^2} &\equiv 0 \end{aligned}\right\} \quad t \in [0,T]$$

or: <u>**Built-in end (fixed support)**</u>

(2.25b)

$$\left.\begin{aligned} y(b,t) &\equiv 0 \\ \frac{\partial y(b,t)}{\partial x} &\equiv 0 \\ \theta(b,t) &\equiv 0 \\ \frac{\partial \theta(b,t)}{\partial x} &\equiv 0 \end{aligned}\right\} \quad t \in [0,T]$$

and the initial conditions

$$\theta(0,x) = \theta_0(x)$$

$$y(0,x) = y_0(x)$$

$$\frac{\partial \theta}{\partial t}(0,x) = \omega_0(x)$$

$$\frac{\partial y(0,x)}{\partial t} = v_0(x),$$

where θ_0, y_0, ω_0, v_0 are continuously differentiable functions of x, whose second derivatives are square integrable, bounded functions on the interval $[0,\ell]$.

2.11.3. The basic control problem

We pose the following control, or excitation problem: Consider the system of equations associated with equations (2.22a), (2.22b):

$$L_1(y,\theta) = \phi_1(x)u_1(t), \tag{2.26a}$$

$$L_2(y,\theta) = \phi_2(x)u_2(t). \tag{2.26b}$$

Where $\phi_i(x)$, $i = 1,2$, is either a bounded piecewise continuous function, or is the Dirac delta function, or its first derivative. In addition we postulate

$$\int_0^\ell |\phi_i| \, dx = 1 , \quad i = 1,2.$$

u(t) will be called an admissible control if $u_i(t)$ are piecewise continuous bounded functions on a time interval $[0,T]$, obeying $|u_i(t)| \leq 1$ for all $t \in [0,T]$. $\phi_i(x)$ is suppose to be given a priori, $\underset{\sim}{u}(t)$ is to be determined. Our problem is to find an admissible control $\underset{\sim}{u}(t)$ such that at the given time $t = T > 0$ the total energy of the beam $\mathcal{E}(T)$

assumes the lowest (greatest) possible value. It is convenient to rewrite the equations (2.26a), (2.26b) in the vector form

$$(2.27) \qquad L(\underset{\sim}{Y}) = \phi\underset{\sim}{u}$$

where $\underset{\sim}{Y}$ stands for $\underset{\sim}{Y} = \begin{bmatrix} y \\ \theta \end{bmatrix}$,

$$\underset{\sim}{u} = \begin{bmatrix} u_1 \\ u_2 \end{bmatrix},$$

$$\phi = \begin{bmatrix} \phi_1 & 0 \\ 0 & \phi_2 \end{bmatrix}.$$

We proceed as in [7] by introducing an inner product < , >.

2.11.4. The energy product

Let $\underset{\sim}{Y}_{(1)}$, $\underset{\sim}{Y}_{(2)}$ be two solutions of (2.27) corresponding to control vectors $\underset{\sim}{u}_{(1)}$, $\underset{\sim}{u}_{(2)}$ respectively.

We define:

$$\langle \underset{\sim}{Y}_{(1)}, \underset{\sim}{Y}_{(2)} \rangle = \frac{\rho}{2} \{ \int_0^{\ell} [A(x) \left(\frac{\partial y_{g_1}}{\partial t}\right)\left(\frac{\partial y_{g_2}}{\partial t}\right)$$

$$+ I_p \left(\frac{\partial \theta_1}{\partial t}\right)\left(\frac{\partial \theta_2}{\partial t}\right)]dx$$

$$+ \frac{E}{2} \int_0^\ell [I(x) \left(\frac{\partial^2 y_1}{\partial x^2}\right)\left(\frac{\partial^2 y_2}{\partial x^2}\right) + C_w(x) \left(\frac{\partial^2 \theta_1}{\partial x^2}\right)\left(\frac{\partial^2 \theta_2}{\partial x^2}\right)]dx$$

$$+ \frac{G}{2} \int_0^\ell C(x)\left(\frac{\partial \theta_1}{\partial x}\right)\left(\frac{\partial \theta_2}{\partial x}\right) dx).$$

(Note that in our case A, I, I_p, C, C_w are constant, and can be brought outside the integral sign.) It is easily checked that this product satisfies all axiomatic requirements of an inner product.

As a consequence we have the Cauchy-Schwartz inequality:

$$\text{(2.28)} \qquad \langle \underset{\sim}{Y}_{(1)}, \underset{\sim}{Y}_{(2)} \rangle^2 \leq \langle \underset{\sim}{Y}_{(1)}, \underset{\sim}{Y}_{(1)} \rangle \cdot \langle \underset{\sim}{Y}_{(2)}, \underset{\sim}{Y}_{(2)} \rangle$$

$$= \mathcal{E}_1(\underset{\sim}{Y}_{(1)})\ \mathcal{E}_2(\underset{\sim}{Y}_{(2)})$$

where $\mathcal{E}_i(\underset{\sim}{Y}_{(i)}) = \mathcal{E}_i(\underset{\sim}{Y}_{(i)}(x,t),t)$, $i = 1,2$, is the total energy corresponding to the vector $\underset{\sim}{Y}_{(i)}$, $i = 1,2$. The equality holds only if for some constant C: $\underset{\sim}{Y}_{(1)} = C\underset{\sim}{Y}_{(2)}$. We compute the time derivative of the product $\langle \underset{\sim}{Y}_{(1)}, \underset{\sim}{Y}_{(2)} \rangle$:

$$\frac{d}{dt} \langle \underset{\sim}{Y}_{(1)}, \underset{\sim}{Y}_{(2)} \rangle = \sum_{i \neq j}^{2} \{\frac{\rho}{2} \int_0^\ell [A(x) \frac{\partial y_{g_i}}{\partial t} \frac{\partial^2 y_{g_j}}{\partial t^2}$$

$$+ I_p \frac{\partial \theta_i}{\partial t} \frac{\partial^2 \theta_j}{\partial t^2}]dx + \frac{E}{2} \int_0^\ell [I(x)$$

$$\frac{\partial^2 y_i}{\partial x^2} \cdot \frac{\partial}{\partial t} \left(\frac{\partial^2 y_j}{\partial x^2} \right) + C_w(x) \frac{\partial^2 \theta_i}{\partial x^2} \frac{\partial}{\partial t} \frac{\partial^2 \theta_j}{\partial x^2})]dx$$

$$+ \frac{G}{2} \int_0^{\ell} [C(x) \frac{\partial \theta_i}{\partial x} \frac{\partial}{\partial t} (\frac{\partial \theta_j}{\partial x})]dx$$

substituting $y_g = y_i - e\theta_i$, we have after integration by parts (and using boundary conditions of either (2.25a), (2.25b))

$$\frac{d}{dt} \langle \underset{\sim}{Y}_{(1)}, \underset{\sim}{Y}_{(2)} \rangle = \sum_{\substack{i,j=1,2 \\ i \neq j}} \{ \frac{1}{2} \int_0^{\ell} \dot{y}_i [\rho A \ddot{y}_j - \rho A e \ddot{\theta}_j$$

$$+ EI \frac{\partial^4 y_j}{\partial x^4}]dx + \frac{1}{2} \int_0^{\ell} \dot{\theta}_i [-\rho I_0 \ddot{\theta}_j - \rho A e \ddot{y}_j$$

$$- GC \frac{\partial^2 \theta_j}{\partial x^2} + EC_w \frac{\partial^4 \theta_j}{\partial x^4}]dx \}$$

$$= \frac{1}{2} \int_0^{\ell} [\dot{y}_1 L_1(y_2, \theta_2) + \dot{y}_2 L_1(y_1, \theta_1)$$

$$+ \dot{\theta}_1 L_2(y_2, \theta_2) + \dot{\theta}_2 L_2(y_1, \theta_1)]dx.$$

$$(2.29) \qquad = \frac{1}{2} \int_0^{\ell} [\dot{\underset{\sim}{Y}}_{(1)} \cdot \underset{\sim}{\Phi}_{(2)} \underset{\sim}{u}_{(2)} + \dot{\underset{\sim}{Y}}_{(2)} \cdot \underset{\sim}{\Phi}_{(1)} \underset{\sim}{u}_{(1)}]dx$$

(The dots stand for differentiation with respect to time). Written out in full the equation (2.29) is:

$$\frac{d}{dt}\langle \underset{\sim}{Y}_{(1)}, \underset{\sim}{Y}_{(2)}\rangle = \frac{1}{2}\int_0^{\ell} \dot{y}_1(x,t)\phi_{1(2)}(x)u_{1(2)}(x)$$

$$+ \dot{\theta}_1(x,t)\phi_{2(2)}(x)u_{2(2)}(t)$$

$$+ \dot{y}_2(x,t)\phi_{1(1)}(x)u_{1(1)}(t)$$

$$+ \dot{\theta}_2(x,t)\phi_{2(1)}(x)u_{2(1)}(t)]dx. \tag{2.29a}$$

The bracketed subscripts referred to the corresponding vectors $\underset{\sim}{Y}_{(1)}$ or $\underset{\sim}{Y}_{(2)}$ respectively. Now let $\underset{\sim}{Y}_{(2)} = \underset{\sim}{Y}_H$ be any solution of the homogeneous equation corresponding to the control vector $\underset{\sim}{u}_{(2)} \equiv 0$. Then

$$\frac{d}{dt}\langle \underset{\sim}{Y}_{(1)}, \underset{\sim}{Y}_H\rangle = \frac{1}{2}\int_0^{\ell} [\dot{y}_H\phi_{1(1)}u_{1(1)}$$

$$+ \dot{\theta}_H\phi_{2(1)}u_{2(1)}]dx,$$

and

$$\langle \underset{\sim}{Y}_{(1)}, \underset{\sim}{Y}_H\rangle_{t=\tau} = \langle \underset{\sim}{Y}_{(1)}, \underset{\sim}{Y}_H\rangle_{t=0}$$

$$+ \frac{1}{2}\int_0^{\tau}\int_0^{\ell} [\dot{y}_H\phi_{1(1)}u_{1(1)}$$

(2.30) $$+ \theta_H \phi_{2(1)} u_{2(1)}]dxdt.$$

2.11.5 The convexity of attainable solutions

LEMMA 1. *Let* $\underset{\sim}{u}_1, \underset{\sim}{u}_2$ *be admissible controls and* $\underset{\sim}{Y}_{(1)} = \begin{bmatrix} Y_1 \\ \theta_1 \end{bmatrix}$, $\underset{\sim}{Y}_{(2)} = \begin{bmatrix} Y_2 \\ \theta_2 \end{bmatrix}$ *be the corresponding attainable solutions. Then* $\hat{\underset{\sim}{Y}} = \lambda \underset{\sim}{Y}_{(1)} + (1-\lambda)\underset{\sim}{Y}_{(2)}$ *is also an attainable solution, for any constant* $0 \leq \lambda \leq 1$.

Proof. First we have to show that the control vector $\hat{\underset{\sim}{u}} = \lambda \underset{\sim}{u} + (1-\lambda)\underset{\sim}{u}_2$ is an admissible control. This follows from our definition and from the triangular inequality:

$$||\hat{\underset{\sim}{u}}|| \leq ||\lambda \underset{\sim}{u}_1|| + ||(1-\lambda)\underset{\sim}{u}_2||$$

$$= |\lambda| \; ||\underset{\sim}{u}_1|| \;\; |1-\lambda| \; ||\underset{\sim}{u}_2|| \leq 1.$$

We now use Duhamel's principle, which postulates the existence of a 2 × 2 kernel matrix $G(x,\xi,t,\tau)$, depending upon the equations (5) and upon the boundary conditions, but independent of the initial conditions, or of the inhomogeneous term, and such that

(2.31) $$\dot{\underset{\sim}{Y}}(x,t) = Y_H(x,t) + \int_0^{\tau} \int_0^{\ell} \underset{\sim}{G}(x,\xi,t,\tau)\underset{\sim}{\phi}(\xi)\underset{\sim}{u}(\tau)d\xi d\tau,$$

where Y_H is the solution of the homogeneous equation obeying the same initial conditions. It follows immediately from (2.31) that the solution $\hat{Y}(x,t) = \lambda Y_1 + (1 - \lambda)Y_2$ will result from the admissible control $\hat{u} = \lambda u_1 + (1 - \lambda)u_2$, hence $\hat{Y}$ is an attainable solution as was to be proved.

2.11.6. The uniqueness of final state due to an optimal control

Theorem 1. Let $\tilde{u}_{(1)}$, $\tilde{u}_{(2)}$ be two optimal controls (that is admissible controls reducing the total energy $\mathcal{E}(t)$ to the lowest possible level $\mathcal{E}_{min}$ at the time $t = T$), and let $\tilde{Y}_{(1)}(x,t)$, $\tilde{Y}_{(2)}(x,t)$ be the corresponding optimal displacement functions. Then $\tilde{Y}_{(1)}(x,T) = \tilde{Y}_{(2)}(x,T)$, and $\dot{\tilde{Y}}_{(1)}(x,T)$, $\dot{\tilde{Y}}_{(2)}(x,T)$ that is the final state determined by any optimal control is unique.

Proof. Let us assume to the contrary that either $\tilde{Y}_{(1)}(x,T) \neq \tilde{Y}_{(2)}(x,T)$, or $\dot{\tilde{Y}}_{(1)}(x,T) \neq \dot{\tilde{Y}}_{(2)}(x,T)$. By the result of lemma 1 the displacement vector $\frac{1}{2}(\tilde{Y}_{(1)} + \tilde{Y}_{(2)}) = \hat{Y}$ is also an attainable solution on the time interval $[0,T]$. (This solution would correspond to the control $\frac{1}{2}(\tilde{u}_{(1)} + \tilde{u}_{(2)})$). The total energy $\mathcal{E}(Y(x,t);t)$ is given by: $\mathcal{E}(\hat{Y}(x,t);t) =$

$$\frac{E}{2}\int_0^{\ell} I\left(\frac{\partial^2 y}{\partial x^2}\right)^2 dx + \frac{G}{2}\int_0^{\ell} C\left(\frac{\partial^2 \theta}{\partial x^2}\right)^2$$

$$+ \frac{E}{2}\int_0^{\ell} C_w\left(\frac{\partial^2 \hat{\theta}}{\partial x^2}\right)^2 dx + \frac{\rho}{2}\int_0^{\ell} A\left(\frac{\partial \dot{y}_g}{\partial t}\right)^2$$

$$+ \frac{\rho}{2} \int_0^{\ell} I_p \left(\frac{\partial \hat{\theta}}{\partial t} \right)^2 dx = \frac{1}{4} \mathcal{E}(\tilde{Y}_{(1)}(x,t);t)$$

$$+ \frac{1}{4} \mathcal{E}(\tilde{Y}_{(2)}(x,t);t) + \frac{1}{2} \langle \tilde{Y}_{(1)}, \tilde{Y}_{(2)} \rangle$$

By Cauchy-Schwartz inequality (which is a strict inequality if $Y_{(1)} \neq Y_{(2)}$, or $\dot{Y}_{(1)} \neq \dot{Y}_{(2)}$:

$$\mathcal{E}(\hat{Y}(x,t),t) < \frac{1}{4} \mathcal{E}(\tilde{Y}_{(1)}(x,t);t)$$

$$+ \frac{1}{4} \mathcal{E}(\tilde{Y}_{(2)}(x,t);t)$$

$$+ \frac{1}{2} \sqrt{\mathcal{E}(\tilde{Y}_{(1)};t)\, \mathcal{E}(\tilde{Y}_{(2)};t)} \tag{2.32}$$

At t = T we have however,

$$\mathcal{E}(\tilde{Y}_{(1)}(x,t);T) = \mathcal{E}(\tilde{Y}_{(2)}(x,t);T) = \mathcal{E}_{min}.$$

Since either $\tilde{Y}_{(1)}(T) \neq \tilde{Y}_{(2)}(T)$ or $\dot{\tilde{Y}}_{(1)}(T) \neq \dot{\tilde{Y}}_{(2)}(T)$ the strict inequality is true in (2.32) and

$$\mathcal{E}(\hat{Y}(x,t),T) < \frac{1}{4} \mathcal{E}_{min} + \frac{1}{4} \mathcal{E}_{min} + \frac{1}{2} \mathcal{E}_{min} = \mathcal{E}_{min},$$

which is a contradiction, since by assumption $\mathcal{E}_{min}$ was the lowest energy level attainable at the time t = T by the use of an admissible control. We note that this line of argument will fail in the case of optimal excitation. If $\underset{\sim}{v}_1(t)$ is an optimal excitation and $\underset{\sim}{v}_2(t)$ is another optimal excitation such that $(\tilde{\underset{\sim}{v}}_1(t),T) = (\tilde{\underset{\sim}{v}}_2(t),T) = \max_{u \in U} \mathcal{S}(u(t),T)$ then $\lambda \underset{\sim}{v}_1 + (1-\lambda)\underset{\sim}{v}_2$ cannot be an optimal excitation unless $\underset{\sim}{Y}(\tilde{\underset{\sim}{v}}_1(t),x,t)_{t=T} = \underset{\sim}{Y}(\tilde{\underset{\sim}{v}}_2(t),x,t)_{t=T}$ and $\dot{\underset{\sim}{Y}}(\tilde{\underset{\sim}{v}}_1(t),x,t)_{t=T} = \dot{\underset{\sim}{Y}}(\underset{\sim}{v}_2(t),x,t)_{t=T}$. The argument supporting this statement is a repetition of the arguments already given above.

2.11.7. Pontryagin's principle

Theorem 2. Let $\underset{\sim}{u} = \begin{bmatrix} u_1(t) \\ u_2(t) \end{bmatrix}$ be the optimal control vector for the fixed interval problem $t \in [0,T]$. Let $\min_{u \in U} \mathcal{S}(T) > 0$. Let $\tilde{\underset{\sim}{W}}_H (\tilde{\underset{\sim}{W}}_H = \begin{bmatrix} y_H \\ \theta_H \end{bmatrix})$ denote the solution of the homogeneous equation satisfying the same final conditions as the optimal solution $\begin{bmatrix} y(\tilde{u}) \\ \theta(\tilde{u}) \end{bmatrix}$. This is

$$y(\tilde{u}(t),x,T) = y_H(x,T)$$
$$\dot{y}(\tilde{u}(t),x,T) = \dot{y}_H(x,T)$$
$$\theta(\tilde{u}(t),x,T) = \theta_H(x,T)$$
$$\dot{\theta}(\tilde{u}(t),x,T) = \dot{\theta}_H(x,T).$$

Then for all $t \in [0,T]$ we have the inequality:

$$- \int_0^{\ell} [\dot{y}_H \phi_1(x) \tilde{u}_1(t) + \dot{\theta}_H \phi_2(x) \tilde{u}_2(t)] dx$$

$$= \max_{u \in U} \{- \int_0^{\ell} [\dot{y}_H \phi_1(x) u_1(t) + \dot{\theta}_H \phi_2(x) u_2(t)] dx\}$$

The proof is lengthy but follows exactly the line of argument used by the author in [17]. For this reason the proof is omitted.

The uniqueness of optimal control depends entirely on the properties of the norm we assign on U, and has little to do with the properties of the linear differential operator L. To state this result we define the concept of strict convexity of the norm $|| \; ||$. The norm $|| \; ||$ is strictly convex (or we say that the unit ball of U is strictly convex), if for any two points u_1, u_2 in U (the closure of U) such that $||u_1|| = ||u_2|| = 1$ the open line segment connecting u_1, u_2, that is the set $\ell = \{z | z = \lambda u_1 + (1-\lambda) u_2, \; 0 < \lambda < 1)\}$ contains only points of norm strictly less than one; i.e. $z \in \ell$ implies $||z|| < 1$. In other words any point lying in the interior of the line segment connecting the boundary points u_1, u_2 of the unit ball lies entirely in the interior of the unit ball. Hence, the unit ball is convex, and contains no straight line segments on its boundary. For example if the unit ball is defined by the Euclidean norm:

$$||u|| < 1 \quad (u_1^2 + u_2^2 + \ldots + u_m^2) < 1,$$

the boundary of the unit ball is a sphere in the m-dimensional Euclidean space, and obviously the unit ball is strictly convex. However, if we assign L_1 norm to U, or the sup. norm (i.e. ℓ_∞ norm) it is convex, but not strictly convex.

2.11.8. The necessary condition for the uniqueness of an optimal control

Let the set of admissible controls U be a dense subset of Banach space B with a norm $|| \; ||$ such that the unit ball of B is strictly convex. Let $\tilde{u}(t)$ be an optimal control which reduces the total energy to the level $\delta_{min} > 0$ at the time $t = T$. Then $\tilde{u}(t)$ is essentially unique in the following sense: any other optimal control $\tilde{u}_1(t)$ can differ from $\tilde{u}(t)$ only by a bounded function which fails to be equal to zero only on a set of measure zero.

Proof

Let us assume that the unit ball of B is strictly convex. By the way of contradiction let us assume that two optimal controls exist $\tilde{\underset{\sim}{u}}_1$, $\tilde{\underset{\sim}{u}}_2$ with

$$-\int_0^\ell (\dot{\underset{\sim}{Y}}_H \phi \tilde{\underset{\sim}{u}}_1)dx = -\int_0^\ell (\dot{\underset{\sim}{Y}}_H \phi \tilde{\underset{\sim}{u}}_2)dx = C = \max_{u \in U} \{-\int_0^\ell (\underset{\sim}{Y}_H \phi \underset{\sim}{u})dx\},$$

such that on some subset M of positive measure of [0,T]

$$\int_M |u_1(t) - u_2(t)|dt > 0.$$

We consider the control:

$$u_w = \frac{1}{2}(\tilde{u}_1 + \tilde{u}_2). \quad \text{We have:}$$

$$(u_w, -\dot{Y}_H) = \int_0^\ell (-Y_H, \phi u_w)dx$$

$$= \frac{1}{2}(\tilde{u}_1, -Y_H) + \frac{1}{2}(\tilde{u}_2, -Y_H)$$

$$= \frac{1}{2}C + \frac{1}{2}C$$

$$= C.$$

But either $\tilde{u}_1$, $\tilde{u}_2$ have a disjoint support, which is impossible, or $\frac{1}{2}||\tilde{u}_1 + \tilde{u}_2|| < \frac{1}{2}\left(||u_1|| + ||u_2||\right)$ because of the strict convexity of the unit ball of B. Hence there exists a number $a > 1$, such that $||au_w|| = a \cdot ||\frac{1}{2}(u_1 + u_2)|| = 1$.

au_w is therefore an admissible control vector. But

$$(au_w, (-\dot{Y}_H)) = a(u_w, (-\dot{Y}_H)) = aC > C$$

which is impossible by the maximality of C. This proves the

uniqueness of $\tilde{u}$. Conversely if we assume that the unit ball of B is not strictly convex, then it is easy to give examples of non-unique controls. (It may be possible to find a control $\hat{u}(t)$, $||\hat{u}|| > 0$ such that on some subinterval of [0,T] we have:

$$\int_0^{\ell} (\dot{Y}_H \phi \hat{u})\,dx = 0,$$

and such that $||\tilde{u} + \hat{u}|| = ||\tilde{u}|| = 1$, if the unit ball of U contains a line segment on its boundary.)

2.11.9. Applications of the maximal principle

It appears that to find the optimal control $\tilde{u}(t)$, we need to know $Y_H(T)$ due to the optimal control. However, an iterative scheme can be immediately suggested on the lines of [17], whereby an arbitrary control could be improved by the use of the maximal principle.

We note that if $\phi \cdot u$ is known then the problem of finding a solution of the coupled vibration can be easily solved numerically. See for example Myklestad [12], or Ker Wilson [13], or by a straightforward application of Rayleigh-Ritz, or Galerkin techniques. Hence, for a known control $\phi(x)u^0(t)Y_H^0(T)$ can be obtained numerically, and we can find the numerical solution $Y_H(x,t)$, $t \in [0,T]$, of the beam vibrating freely "backward" in time. Now the maximality principle can be used to improve our control. Using this new control $\phi(x)u^1(t)$, we can obtain $\underset{\sim}{Y}_H^{(1)}(T)$ and repeat the entire computation.

This is obviously tedious, and no assurance can be given that such improved sequence of controls converges to an optimal control. These problems are easily overcome if we become less ambitious in our definition of optimality. If following [17] we define "optimal" to be instantly optimal in the sense used by the author, then the optimality test can be applied immediately, <u>without first determining the final state</u>. In fact an analogous argument gives us this important theorem:

<u>Theorem 4</u>. The instantly optimal control $\hat{u}(t)$ satisfies the maximality principle:

$$\int_0^\ell \dot{\underset{\sim}{Y}}(\hat{u}(t),x,t)\phi(x)\cdot\hat{\underset{\sim}{u}}(t)dx$$

$$= \max_{u\varepsilon U} \{-\int_0^\ell \dot{\underset{\sim}{Y}}(u(t),x,t)\phi(x)\underset{\sim}{u}(t)dx$$

for all $t \varepsilon [0,T]$.

The definition of instantly optimal is analogous to the one given in [17]. The instantly optimal control coincides with the control reducing the energy at every instant in [0,T] at the maximum possible rate. Again such control is easily shown to be unique if $||\cdot||_m$ defined on U is strictly convex.

APPENDIX 2.1

Formulas for Torsion Constant C and Warping Constant C_w for some cross-sections.

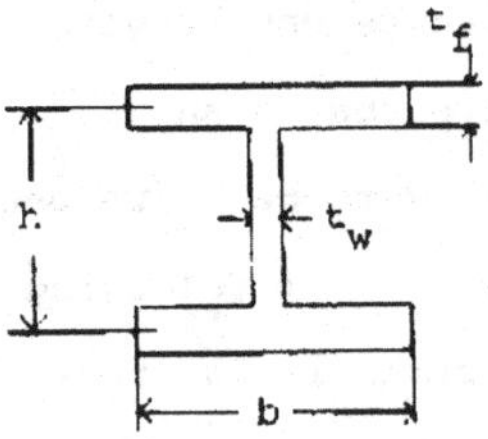

$$\begin{cases} C = \dfrac{h t_w^3 + 2 b t_f^3}{3} \\[2ex] C_w = \dfrac{t_f h^2 b^3}{24} \end{cases}$$

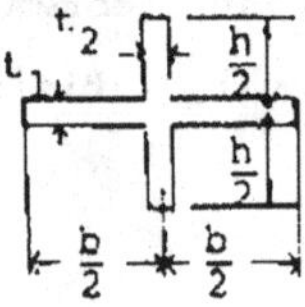

$$\begin{cases} C = \dfrac{h t_2^3 + b t_1^3}{3} \\[2ex] C_w = 0 \end{cases}$$

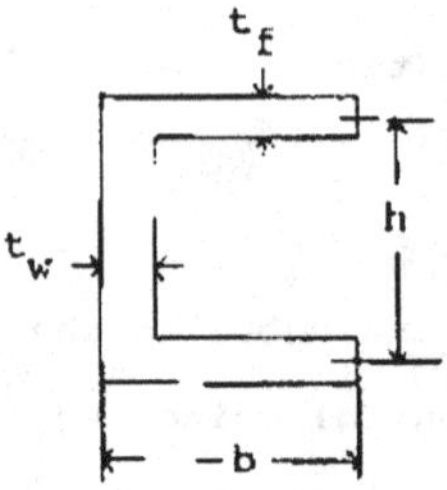

$$\begin{cases} C = \dfrac{h t_w^3 + 2 b t_f^3}{3} \\[2ex] C_w = \dfrac{t_f h^2 b^3 (2h t_w + 3b t_f)}{h t_w + 6 b t_f} \end{cases}$$

CHAPTER III

Optimal Control Theory for Thin Plates

3.0 Introductory Comments

This chapter establishes Pontryagin's principle for thin plates subject to some physically motivated boundary conditions. Instantly optimal control is also studied in this chapter. Finally some comments are made concerning optimal excitation problems in the thin plate theory. Some results obtained in this chapter are a direct generalization of the results given in the preceding chapter for the one-dimensional case. However, it becomes clear that in most cases the obvious generalization of the beam theory does not succeed, and that Pontryagin's maximal principle becomes very complex.

3.1.0. Assumptions and Notation

We assume the usual linear hypothesis of thin plate theory listed below and numbered (i)-(iv), plus "elastica" assumptions listed as (E1)-(E4). These assumptions imply Duhamel's principle as given in the paragraph (3.1.8). The plate is assumed to occupy a compact simply connected region $\bar{\Omega}$ of the Euclidean space E^2. The interior of $\bar{\Omega}$ will be denoted by Ω, and the boundary of $\bar{\Omega}$ by $\partial\Omega$. The boundary $\partial\Omega$ consists of smooth

Jordan curves. The problem of corner points will be considered only in the final paragraphs of Section 2 of this chapter. Prior to that the boundary curves will be assumed smooth (i.e. of the class C^1).

Notation and the physical meaning of symbols used here is as follows:

x,y,z will denote Cartesian coordinates of E^3.

t will denote time.

The plate occupies a region $\Omega \subset E^2$, the plane of Ω being spanned by the coordinates x, y.

u,v,w may denote the displacements in the directions of the axes x,y,z respectively.

*v-transverse velocity $v = \frac{dw}{dt}$.

E - Young's modulus ($E > 0$).

ν - Poisson's ratio, ($0 < \nu < \frac{1}{2}$).

h - thickness of the plate, ($h > 0$ in Ω).

D - flexual rigidity: $D = \dfrac{Eh^3}{12(1-\nu^2)}$.

ρ - the mass density (mass per unit area).

$\partial\Omega$ - the boundary of Ω.

n - the unit vector in the direction of the outward normal to $\partial\Omega$.

τ - the unit vector in the direction tangential to $\partial\Omega$.

ds - increment of length.

---

*despite of the two different meanings assigned to the symbol v, no confusion will arise, since only the displacement $w(x,y,t)$ will be considered significant in the simplified theory considered in this monograph.

$\epsilon_{xx}, \epsilon_{xy}, \epsilon_{yy}$ - the linear strains.

$\tau_{xx}, \tau_{xy}, \tau_{yy}$ - the linear stresses. (The stress system is assumed to be two-dimensional).

M_{ij} - the moments.

N_{ij} - the normal forces (per unit length).

Q_i - the shear forces (per unit length).

T - Kinetic energy.

U - Strain energy.

$\mathcal{E}$ - total energy. ($\mathcal{E} = T + U$).

The displacement functions u(x,y,t), v(x,y,t), w(x,y,t) shall satisfy the "elastica" hypothesis listed below:

(E1) For a fixed t the functions u,v,w and their time derivatives $\frac{\partial u}{\partial t}, \frac{\partial v}{\partial t}, \frac{\partial w}{\partial t}$ possess in Ω the partial derivatives of order one and two with respect to x and y.

(E2) The two dimensional strain components are twice differentiable functions almost everywhere in Ω.

(E3) For each fixed x and y the displacement function w(x,y,t) is a continuously differentiable function of t.

(E4) The functions $D(x,y)(\nabla^2 w)^2 - (1-\nu)\diamondsuit^4(w,w)$ and $\rho(x,y)h(x,y)\frac{\partial w}{\partial t}$ are bounded and measurable functions of t and are square integrable in Ω (for every fixed value of $t \in [0,\infty]$). ∇^2 denotes the Laplace operator, $\diamondsuit^4$ is defined by the formula (13).

The thin plate linear theory

In this part we shall consider the control, or excitation of plates, subject to the classical thin plate, small deflection theory, as proposed by Lagrange. A detailed discussion is given in [22]. The assumptions leading to this simplified model are

i) Hooke's law

ii) The slope of the deflected plate, taken in an arbitrary direction is sufficiently small so that its square can be neglected when compared with unity.

iii) The mid-plane of the plate is the neutral plane of the deflected plate, that is the direct stresses $\tau_{xx}, \tau_{yy}, \tau_{zz}$ acting on the mid-plane are neglected.

iv) Stress components normal to the mid-plane of the plate can be ignored.

v) A line normal to the mid-plane of the undeflected plate will remain normal to the mid-plane of the deflected plate. Some alternate large deflection theories shall be considered in later developments of this theory.

3.1 A brief outline of the small deflection theory

3.1.1. The Basic Equations

The assumptions (ii)-(v) suggest that the plate may be

considered to be a subset of the Euclidean plane E^2, and the displacements u,v in the directions of the Cartesian coordinates x and y respectively (of E^2) are linearly varying with the distance z from the mid-plane of the plate:

$$u = -z \frac{\partial w}{\partial x} \qquad (3.1^a)$$

$$v = -z \frac{\partial w}{\partial y} \qquad (3.1^b)$$

We shall now review the formal development of the small deflection theory. w is the deflection of the plate, i.e., displacement in the direction of the z-axis. The strain components are given by the usual linear approximations:

$$\epsilon_{xx} = \frac{\partial u}{\partial x} = -z \frac{\partial^2 w}{\partial x^2} \qquad (3.2^a)$$

$$\epsilon_{xy} = \epsilon_{yx} = \frac{\partial u}{\partial y} + \frac{\partial v}{\partial x} = -2z \frac{\partial^2 w}{\partial x \partial y} \qquad (3.2^b)$$

$$\epsilon_{yy} = \frac{\partial v}{\partial y} = -z \frac{\partial^2 w}{\partial y^2} \qquad (3.2^c)$$

Following the assumption (iii) the effects of the strain components $\epsilon_{xz}, \epsilon_{yz}, \epsilon_{zz}$ are going to be ignored.

We assume the correctness of Hooke's law:

$$\epsilon_{xx} = \frac{1}{E} (\tau_{xx} - \nu\tau_{yy}) \qquad (3.3^a)$$

$$\varepsilon_{xy} = \frac{2(1+\nu)}{E} \tau_{xy} \qquad (3.3^b)$$

$$\varepsilon_{yy} = \frac{1}{E} (\tau_{yy} - \nu\tau_{xx}) \qquad (3.3^c)$$

where E denotes the Young's modulus, and ν denotes the Poisson's ratio. Solving for $\tau_{xx}, \tau_{xy}, \tau_{yy}$ the equations (3.3^a), (3.3^b) and (3.3^c) and substituting (3.2^a), (3.2^b), and (3.2^c) gives the formulae:

$$\tau_{xx} = - \frac{Ez}{1-\nu^2} \left(\frac{\partial^2 w}{\partial x^2} + \nu \frac{\partial^2 w}{\partial y^2} \right) \qquad (3.4^a)$$

$$\tau_{xy} = \tau_{yx} = - \frac{Ez}{1+\nu} \frac{\partial^2 w}{\partial x \partial y} \qquad (3.4^b)$$

$$\tau_{yy} = - \frac{Ez}{1-\nu^2} \left(\frac{\partial^2 w}{\partial y^2} + \nu \frac{\partial^2 w}{\partial x^2} \right) \qquad (3.4^c)$$

The moments acting on the plate are given by the formulae:

$$M_{xx} = \int_{-h/2}^{+h/2} z\tau_{xx} dz = -D\left(\frac{\partial^2 w}{\partial x^2} + \nu \frac{\partial^2 w}{\partial y^2}\right) \qquad (3.5^a)$$

$$M_{xy} = -M_{yx} = \int_{-h/2}^{+h/2} z\tau_{xy} dx = -D(1-\nu) \frac{\partial^2 w}{\partial x \partial y} \qquad (3.5^b)$$

$$M_{yy} = \int_{-h/2}^{+h/2} z\tau_{yy} ds = -D\left(\frac{\partial^2 w}{\partial y^2} + \nu \frac{\partial^2 w}{\partial x^2}\right) , \qquad (3.5^c)$$

where h denotes the thickness of the plate, and where D is the flexual rigidity of the plate defined by the relationship:

$$D = \frac{Eh^3}{12(1-\nu^2)} \quad . \tag{3.6}$$

The shear forces (i.e. forces normal to the plane of the plate) are expressed in terms of moments as follows:

$$Q_x = \frac{\partial M_{xx}}{\partial x} + \frac{\partial M_{xy}}{\partial y} \tag{3.7^a}$$

$$Q_y = - \frac{\partial M_{yx}}{\partial x} + \frac{\partial M_{yy}}{\partial y} \tag{3.7^b}$$

Using the expressions (3.5^a), (3.5^b) and (3.7^a) we have:

$$Q_x = \frac{\partial}{\partial x}\left[-D\left(\frac{\partial^2 w}{\partial x^2} + \nu\frac{\partial^2 w}{\partial y^2}\right)\right] + \frac{\partial}{\partial y}\left[-D(1-\nu)\frac{\partial^2 w}{\partial x \partial y}\right]$$

$$= \left(-D\frac{\partial}{\partial x}\nabla^2 w\right) - \frac{\partial D}{\partial x}\left(\frac{\partial^2 w}{\partial x^2} + \nu\frac{\partial^2 w}{\partial y^2}\right) - \frac{\partial D}{\partial y}(1-\nu)\frac{\partial^2 w}{\partial x \partial y} \quad . \tag{3.7^c}$$

Similar formula is obtained for Q_y. If D = const. then (3.7^c) reduces to:

$$Q_\xi = -D\frac{\partial}{\partial \xi}(\nabla^2 w) . \tag{3.7^d}$$

The normal forces, i.e. the forces acting in the plane of the plate will be denoted by N_{xx}, N_{xy}, N_{yy}. (Note the dimension of the forces Q_i, N_{ij}, is force per unit length, i.e. lbs./ft., kg/cm., etc. ...). The equations of equilibrium written in terms of forces are:

$$\frac{\partial Q_x}{\partial x} + \frac{\partial Q_y}{\partial y} + q(x,y) = 0 \qquad (3.8^a)$$

$$\frac{\partial N_{xx}}{\partial x} + \frac{\partial N_{xy}}{\partial y} = 0 \qquad (3.8^b)$$

$$\frac{\partial N_{yy}}{\partial y} + \frac{\partial N_{xy}}{\partial x} = 0 \qquad (3.8^c)$$

where q(x,y) is the applied load (lbs/sq. ft., kg/cm^2, etc. ...).

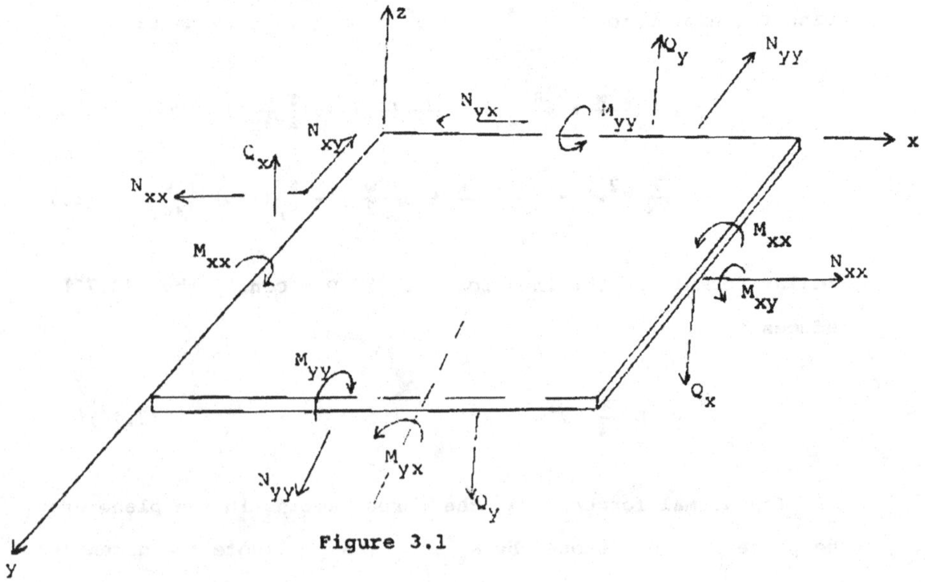

Figure 3.1

The equations of compatibility (see for example [36], page 181, equation (d)) reduce to a simple equation:

$$\frac{\partial^2}{\partial y^2}\left(\frac{\partial u}{\partial x}\right) + \frac{\partial^2}{\partial x^2}\left(\frac{\partial v}{\partial y}\right) - \frac{\partial^2}{\partial x \partial y}\left(\frac{\partial u}{\partial y} + \frac{\partial v}{\partial x}\right) = 0 \qquad (3.9)$$

The equation of compatibility rewritten in the terms of normal forces becomes

$$\frac{\partial^2}{\partial y^2}\left(\frac{N_{xx}-\nu N_{yy}}{h}\right) + \frac{\partial^2}{\partial x^2}\left(\frac{N_{yy}-\nu N_{xx}}{h}\right) - 2(1+\nu)$$

$$\frac{\partial^2}{\partial x \partial y}\left(\frac{N_{xy}}{h}\right) = 0 \quad . \qquad (3.10)$$

(This follows directly from (3.3^a), (3.3^b), (3.3^c), and the assumption (ii).)

Introducing an Airy's stress function ϕ, for the normal forces, twice differentiable in the region $\Omega \subset E^2$ occupied by the plate, and such that

$$\frac{\partial^2 \phi}{\partial x^2} = N_{yy} \qquad (3.11^a)$$

$$\frac{\partial^2 \phi}{\partial x \partial y} = - N_{xy} \qquad (3.11^b)$$

$$\frac{\partial^2 \phi}{\partial y^2} = N_{xx} \quad , \qquad (3.11^c)$$

We see that the equations of equilibrium 3.8^a, 3.8^b, 3.8^c are identically satisfied, and the equation of compatibility becomes

$$\nabla^2\left(\frac{1}{h}\nabla^2\phi\right) - (1+\nu)\diamond^4\left(\frac{1}{h},\phi\right) = 0 \quad , \qquad (3.12)$$

where as before ∇^2 is the Laplace operator, and

$$\diamond^4(A,B) = \frac{\partial^2 A}{\partial x^2}\frac{\partial^2 B}{\partial y^2} - 2\frac{\partial^2 A}{\partial x\partial y}\frac{\partial^2 B}{\partial x\partial y} + \frac{\partial^2 A}{\partial y^2}\frac{\partial^2 B}{\partial x^2} \qquad (3.13)$$

can be regarded as a dot product of the vectors:

$$\left(\frac{\partial^2 A}{\partial x^2}, +\frac{\partial^2 A}{\partial x\partial y}, -\frac{\partial^2 A}{\partial y\partial x}, \frac{\partial^2 A}{\partial y^2}\right),$$

and

$$\left(\frac{\partial^2 B}{\partial y^2}, -\frac{\partial^2 B}{\partial y\partial x}, +\frac{\partial^2 B}{\partial x\partial y}, \frac{\partial^2 B}{\partial x^2}\right),$$

assuming that

$$\frac{\partial^2 A}{\partial x\partial y} = \frac{\partial^2 A}{\partial y\partial x}$$

and

$$\frac{\partial^2 B}{\partial x\partial y} = \frac{\partial^2 B}{\partial y\partial x} .$$

The equation of equilibrium (3.8^a) expressed in terms of the moments becomes

$$\frac{\partial^2 M_{xx}}{\partial x^2} - \frac{\partial^2 M_{xy}}{\partial x \partial y} + \frac{\partial^2 M_{yx}}{\partial x \partial y} + \frac{\partial^2 M_{yy}}{\partial y^2} + q = 0, \qquad (3.14)$$

and substituting formulas (3.5^a), (3.5^b), (3.5^c) we obtain the well-known deflection equation of the small deflection, thin plate theory:

$$\frac{\partial^2}{\partial x^2}\left[D\left(\frac{\partial^2 w}{\partial x^2} + \nu\frac{\partial^2 w}{\partial y^2}\right)\right] + 2(1-\nu)\frac{\partial^2}{\partial x \partial y}\left(D\,\frac{\partial^2 w}{\partial x \partial y}\right)$$

$$+ \frac{\partial^2}{\partial y^2}\left[D\,\frac{\partial^2 w}{\partial y^2} + \nu\,\frac{\partial^2 w}{\partial x^2}\right] - q = 0. \qquad (3.15)$$

This can be rewritten in the form:

$$\nabla^2(D\nabla^2 w) - (1-\nu)\diamond^4(D,w) - q = 0 \qquad (3.15^a)$$

3.1.2. The Case of a Constant Crossection

In this case h = constant = h_o, D = constant = D_o. The equation (3.12) assumes the simplified form

$$\nabla^4\Phi = 0 \quad , \qquad (3.16)$$

while equation (3.15^a) becomes:

$$\nabla^4 w = \frac{q}{D} \, . \tag{3.17}$$

3.1.3 The Basic Dynamic Equations of Small Deflection Theory

Since the nature of the load q has not been specified, we may assume that q is in part the inertia load, opposing the acceleration of the plate, that is the load q consists partially of an outside load $q_o(t)$ and partially of an inertia load $-\rho(x,y)\frac{\partial^2 w}{\partial t^2}$, where $\rho(x,y)$ is the mass density of the plate (i.e., mass per unit area).

The equation (3.15^a) is rewritten in the form:

$$\nabla^2(D\nabla^2 w) - (1-\nu)\Diamond^4(D,w) + \rho\frac{\partial^2 w}{\partial t^2} = q_o + f_N \tag{3.18}$$

where f_N is the vertical load arising from the existence of the normal forces acting on the middle surface of the plate. f_N is computed to be:

$$f_N = N_{xx}\frac{\partial^2 w}{\partial x^2} + 2N_{xy}\frac{\partial^2 w}{\partial x \partial y} + N_{yy}\frac{\partial^2 w}{\partial y^2} = \Diamond^4(\phi,w) \tag{3.19}$$

Combining (3.18) and (3.19), we obtain the basic partial differential system, describing the behavior of a vibrating plate:

$$\nabla^2\left(\frac{1}{h}\nabla^2\phi\right) - (1+\nu)\diamond^4\left(\frac{1}{h},\phi\right) = 0 \qquad (3.A1)$$

$$\rho\frac{\partial^2 w}{\partial t^2} + \nabla^2(D\nabla^2 w) - (1-\nu)\diamond^4(D,w) - \diamond^4(\phi,w) = q \qquad (3.A2)$$

where q(x,y,t) now represents the outside loading imposed on the plate. The dynamic load q(x,y,t) will be called <u>the control function</u>, or simply control. The usual problems of plate theory presuppose the knowledge of q(x,y,t) and seek a solution of the system (3.A1), (3.A2), subject to suitable boundary and initial conditions. Here we shall consider a different type of a problem. We shall seek certain types of solutions, or rather solutions with some property, and pose this question: Subject to the given boundary and initial conditions, can we exhibit a function (or generalized function) q(x,y,t) such that the corresponding solution w(x,y,t) does indeed possess the required property?

In the discussion of solutions of the linear equation (3.A2) we shall need the solutions of the corresponding homogeneous equation:

$$\frac{\partial^2 w}{\partial t^2} + \nabla^2(D\nabla^2 w) - (1-\nu)\diamond^4(D,w) - \diamond^4(\phi,w) = 0 \quad , \qquad (3.A2^a)$$

and we shall consider in particular the solutions of the simpler systems:

$$L(w) = \rho \frac{\partial^2 w}{\partial t^2} + \nabla^2 (D\nabla^2 w) - (1-\nu) \Diamond^4 (D,w) = q, \qquad (3.A3)$$

and of the corresponding homogeneous equation:

$$L(w) = 0. \qquad (3.A3^a)$$

3.1.4. The Boundary Conditions

In the remainder of this work Ω will denote the region occupied by the plate ($\Omega \subset E^2$), and unless it is stated otherwise Ω shall be a connected, compact subset of the Euclidean plane E^2, whose boundary $\partial\Omega$ is a simple, closed Jordan curve. Other properties of Ω will be assumed after an introduction of some basic geometric concepts.

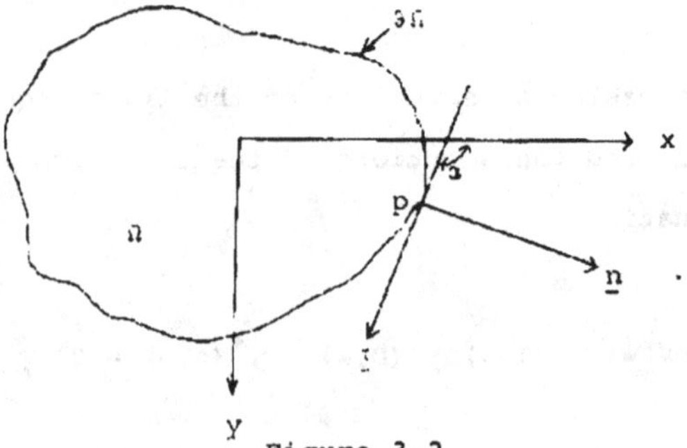

Figure 3.2

Assuming that except for finitely many points of $\partial\Omega$ the outward normal unit vector $\underline{n}$ and the tangential unit vector $\underline{\tau}$ are defined, then selecting a point $p \in \partial\Omega$ at which these directions exist, we define $\frac{\partial}{\partial n}$, $\frac{\partial}{\partial \tau}$ to be the directional derivatives:

$$\frac{\partial f}{\partial n} = \frac{\partial f}{\partial x} \cos (x,n) + \frac{\partial f}{\partial y} \cos (y,n)$$

$$\frac{\partial f}{\partial \tau} = \frac{\partial f}{\partial x} \cos (x,\tau) + \frac{\partial f}{\partial y} \cos (y,\tau).$$

Since the angles between x and n, and between y and n are complementary, we obtain the relationships

$$\frac{\partial^2 w}{\partial \tau^2} = \frac{\partial^2 w}{\partial s^2} + k \frac{\partial w}{\partial n} \qquad (3.20^a)$$

$$\frac{\partial^2 w}{\partial \tau \partial n} = \frac{\partial^2 w}{\partial s \partial n} - k \frac{\partial w}{\partial s} \qquad (3.20^b)$$

where $\frac{\partial}{\partial s}$ denotes the differentiation along the coordinate following the boundary $\partial\Omega$, and k is the curvature of the boundary:

$$k = \frac{\partial \alpha}{\partial s} \qquad (3.20^c)$$

Again we assume here that the second derivatives in (3.20^a),

(3.20^b) are defined on $\partial\Omega$, possibly with the exception of finitely many points. (It is clear that $\frac{\partial w}{\partial \tau} \equiv \frac{dw}{ds}$ whenever this derivative is defined on $\partial\Omega$.)

The boundary conditions will be of either of the three basic types:

<u>The clamped edge</u>:

$$\left.\begin{aligned} W &= 0, \\ \text{and} \qquad \frac{\partial w}{\partial n} &= 0, \end{aligned}\right\} \tag{3.B1}$$

<u>The simply supported edge</u>:

$$\left.\begin{aligned} W &= 0, \\ \frac{\partial^2 w}{\partial n^2} + \nu\frac{\partial^2 w}{\partial \tau^2} &= 0, \end{aligned}\right\} \tag{3.B2}$$

The second condition in (3.B2) may be replaced by:

$$\frac{\partial^2 w}{\partial n^2} + \nu\left(\frac{\partial^2 w}{\partial s^2} + k\,\frac{\partial w}{\partial n}\right) = 0 \tag{3.B2^a}$$

by a substitution of the relationship (3.20^a) in (3.B2).

<u>The free edge</u>:

The absence of external moments, or forces on the free edge would require the simultaneous vanishing of M_{nn}, M_{ns}, and Q_n on the subarc of $\partial\Omega$ which is the free edge. It has been shown by Kirchhoff [22] that these three conditions are equivalent to the pair of conditions:

$$\left.\begin{aligned} M_{nn} &= 0 \\ Q_n + \frac{\partial M_{ns}}{\partial s} &= 0 \end{aligned}\right\} \qquad (3.B3)$$

Substituting the expressions (3.5^a), (3.5^b) and (3.7^a), we can rewrite (3.B3) in the form:

$$\frac{\partial^2 w}{\partial n^2} + \nu\left(\frac{\partial^2 w}{\partial s^2} + k\,\frac{\partial w}{\partial n}\right) = 0$$

$$D\left[\frac{\partial}{\partial n}\,\nabla^2 w + (1-\nu)\,\frac{\partial}{\partial s}\left(\frac{\partial^2 w}{\partial n\partial s} - k\,\frac{\partial w}{\partial s}\right)\right] \qquad (3.B3^a)$$

$$+ \frac{\partial D}{\partial n}\left[\frac{\partial^2 w}{\partial n^2} + \nu\left(\frac{\partial^2 w}{\partial s^2} + k\,\frac{\partial w}{\partial n}\right)\right] + 2(1-\nu)\,\frac{\partial D}{\partial s}$$

$$\left(\frac{\partial^2 w}{\partial n^2} - k\,\frac{\partial w}{\partial s}\right) = 0$$

Some obvious simplifications occur in the formulas (3.B1), $(3.B2^a)$, $(3.B3^a)$ in the cases when either the edge is a straight line (that is $k = 0$) or when the flexural rigidity of the plate $D(x,y)$ is constant in Ω. We recall that in the case $D = $ constant the second formula $(3.B3^a)$ can be rewritten in the form:

$$\frac{\partial}{\partial n}(\nabla^2 w) + (1-\nu)\frac{\partial}{\partial s}\left(\frac{\partial^2 w}{\partial n \partial s} - k\,\frac{\partial w}{\partial s}\right) = 0.$$

In the case when the edge is supported so that the support reacts on the edge with a restoring force f(w), the conditions (3.B3) are modified to read

$$\begin{aligned} M_{nn} &= 0 \\ Q_n + \frac{\partial M_{ns}}{\partial s} + f(w) &= 0 \end{aligned} \qquad (3.B4)$$

In particular if f(w) = Kw where K is a constant, we obtain the usual assumption of an elastic support.

3.1.5. The Initial Conditions

We shall consider the solutions w(x,y,t) of the equations (3.A1) and (3.A2) in the region $\Omega \subset E^2$, for values of $t \geq 0$, where w obeys the conditions (3.B1) and (3.B2) and (3.B3) on subarcs Γ_1, Γ_2, and Γ_3 of $\partial\Omega$, such that $\bar{\Gamma}_1 \cup \bar{\Gamma}_2 \cup \bar{\Gamma}_3 = \partial\Omega$. (Bars over the symbols indicate the closure operation.) w(x,y,t) also obeys the initial conditions

$$w(x,y,0) = \psi(x,y) \qquad (3.C1)$$

$$\frac{\partial w(x,y,0)}{\partial t} = \eta(x,y) \qquad (3.C2)$$

in Ω.

The solution $w(x,y,t)$ is assumed to obey the "elastica" hypothesis, i.e. we shall consider only displacements $w(x,y,t)$ which are continuously differentiable functions of the variables x,y,t in $\Omega \times [0,\infty)$, obeying some additional energy conditions which were stated in (E4). The initial position and velocity functions $\psi(x,y)$, $\eta(x,y)$ are also assumed to obey the "elastica" hypothesis.

3.1.6. The Energy Terms

If only the bending action of the plate is considered, and no energy is supplied or absorbed on the boundary, then following the assumptions (i) - (v), the strain energy of the plate is identified with the complementary energy and is given by:

$$U(t) = \frac{1}{2} \iint_{\Omega} D(x,y) [(\frac{\partial^2 w}{\partial x^2})^2 + (\frac{\partial^2 w}{\partial y^2})^2$$

$$+ 2\nu \frac{\partial^2 w}{\partial x^2} \frac{\partial^2 w}{\partial y^2} + 2(1-\nu)(\frac{\partial^2 w}{\partial x \partial y})^2] dxdy$$

$$= \frac{1}{2} \iint_{\Omega} D[\frac{\partial^2 w}{\partial x^2} + \frac{\partial^2 w}{\partial y^2}]^2 - 2(1-\nu)\{\frac{\partial^2 w}{\partial x^2} \frac{\partial^2 w}{\partial y^2}$$

An additional term: $\frac{1}{2}\iint\limits_{\Omega}\{\frac{1}{2}\;{}^{4}(w^{2},\Phi) - w\diamond^{4}(w,\Phi)\}dxdy$ (3.21[a])

has to be included if the middle forces contribute substantially to the potential energy level.

The kinetic energy of the plate is

$$T = \frac{1}{2}\iint\limits_{\Omega}\rho\left(\frac{\partial w}{\partial t}\right)^{2}dxdy \quad , \tag{3.22}$$

and the total energy of the plate is

$$\mathcal{E}(t) = U(t) + T(t). \tag{3.23}$$

Since no provision was made in the assumptions (i)-(v) for internal dissipation of energy, the total energy is assumed to be constant in the case of a free vibration.

3.1.7. A General Discussion of the Applied Loads

We define the admissible controls in complete analogy with the definition given in Chapter 2. The inequality (2.6) becomes:

$$||q(x,y,t)||^{2} = [\int\limits_{\Omega} |q(x,y,t)dxdy|^{2} \leq 1 \tag{3.24}$$

The equation (2.7) assumes the form:

$$q(x,y,t) = \sum_{i=1}^{N} \delta(x-\xi_{i}(t),y-\eta_{i}(t))\phi_{i}(t)+\sigma(x,y,t). \tag{3.25}$$

with

$$\sum_{i=1}^{N} |\phi_i(t)| + \int_{\Omega} |\sigma(x,y,t)|dxdy \leq 1. \tag{3.26}$$

The remainder of the section dealing with admissible controls for beams can be rewritten with only the obvious alterations. The section (2.2) containing the remarks is also clearly applicable in the two-dimensional case. In particular we retain the definition of the distributed load controls.

Let N_ε denote ε-neighborhood of $\partial\Omega$ for some $\varepsilon > 0$. In analogy with the admissible distributed load control we define the admissible concentrated load (or point load) control to satisfy the inequality:

$$||q||_{\Omega} = \lim\sup \iint_{\Omega-N_\varepsilon} |q(x,y,t)|dxdy \equiv \lim_{\varepsilon>0}\sup (|q(x,y,t)|,\Psi)_{\Omega-N_\varepsilon} \leq 1. \tag{3.26}$$

Here Ψ denotes the test function $\psi(x,y,t) \equiv 1$ for all $x,y \in \Omega-N_\varepsilon(\partial\Omega)$ and any fixed $t \in [0,\infty)$. Point load controls $\psi(x,y,t)$ ($\psi(x,y,t) \equiv 0$ outside of Ω) will be considered of the form $\sum_{i=1}^{N} \delta(x-\xi_i(t),y-\eta_i(t))\phi_i(t)$ where $\xi_i(t)$, $\eta_i(t)$ are functions whose domain is $[0,\infty)$ and the range lies in Ω, i.e. for any $t \in [0,\infty)$ the pair $(\xi_i(t),\eta_i(t))$ can be identified with the x,y coordinates of a point in Ω. In addition the functions $\phi_i(t)$ are bounded and measurable on $[0,\infty)$ obeying:

$$\sum_{i=1}^{N} |\phi_i(t)| \leq 1. \tag{3.27}$$

It should be made clear that we are not interested in the exact description of the control functions, but rather in equivalence classes of such functions. Two control functions will be identified with each other if the resulting displacements are equal. Hence two distributed load controls, whose values differ only on a set of measure zero in $\Omega \times [0,\infty)$ will be regarded as equal.

To summarize our discussion we offer this definition: A bounded linear functional $f(x,y,t)$ over the space of test functions obeying the elastica hypothesis will be called an admissible control, if $||f(x,y,t)||_\Omega \leq 1$; that is, the usual sup. norm of the functional f over its domain is bounded above by 1, and $F(t) = ||f(x,y,t)||_\Omega$ is a measurable function of the time variable t. Since F(t) is measurable and uniformly bounded on the finite interval $[0,\tau]$ it is square integrable and also an absolutely integrable function of t. This immediately implies that the total energy of the plate is also uniformly bounded on $[0,\tau]$ for any positive number τ.

3.1.8. The Statement of Duhamel's Principle

In what follows we shall assume without proof the existence and uniqueness of the solutions of the mixed boundary value and initial value problem (MBVP), i.e. of the problem posed by equations (3.A1), (3.A2) obeyed in $\Omega \times [0,\infty)$ with suitable boundary conditions of the form (3.B1), (3.B2), (3.B3) prescribed on $\partial\Omega$, and with initial conditions (3.C1) and (3.C2), obeying the "elastica" hypothesis, and with an admissible control $q(x,y,t)$ given a priori. Such system will be denoted by MBVP. Rigorous proofs of existence and uniqueness of solutions of MBVP for thin plates has never been offered in the literature to the best knowledge of the author. However, the static case has been given both in [27] (see Chapters 14, 15, and 19) and in a sequence of papers by D. I. Sherman (see for example [34]. The generalization of these proofs to MBVP appear to be straightforward. These proofs, and particularly the existence proofs are hard, and in general the engineering literature dealing with the theory of plates and shells carefully avoids all references to existence and uniqueness of the solutions of the corresponding mathematical systems.

Let $w_H(x,y,t)$ denote the solution of the homogeneous MBVP (3.A1), (3.A2), (i.e. it corresponds to the case when $q(x,y,t) \equiv 0$), with the appropriate boundary value conditions of the form (3.B1), (3.B2), (3.B3), and the initial conditions of the form: (3.C1), (3.C2). Let $q(x,y,t)$ be a distributed load control function. Then the Duhamel's principle asserts that the solution of MBVP: $w(x,y,t)$ will be given by the formula:

$$w(x,y,t) = w_H(x,y,t)$$

$$+ \int_0^t \int_\Omega G((x-\xi),(y-\eta),(t-\tau))q(\xi,\eta,\tau)d\xi d\eta d\tau$$

$$= w_H + G*q \qquad (3.28)$$

where $G(x-\xi,y-\eta,t-\tau)$ depends only on the equations (3.A1), (3.A2) and on the boundary conditions, but does not depend on the initial conditions (3.C1), (3.C2), or on the function $q(x,y,t)$. The symbol * will denote the convolution operation. (See for example [15], Vol.1, page 103.)

In the case where $q(x,y,t)$ is a generalized function of x and y, the double integral has to be interpreted as a bilinear product, as explained in the preceeding paragraph. The proofs of Duhamel's principle for linear systems of differential equations can be found in textbooks. The usual proof (see for example Yosida [24], pages 76-80) utilizes the Ascoli-Arzela theorem, and clearly is not applicable in the case when $q(x,y,t)$ is a point load. However, it is not hard to extend the proof to cover this case as well, by considering the point load to be the limit of a δ-convergent sequence.

3.1.9. Some Identities Arising from the Equation (3.A2) Under the Assumption that the Effects of the Middle Plane Forces are Negligible

Following the above assumption the equation (3.A2) can be rewritten in the simpler form:

$$\nabla^2(D\nabla^2 w) - (1-\nu)\diamond^4(D,w) + \rho\frac{\partial^2 w}{\partial t^2} = q. \qquad (3.A2^a)$$

The strain energy of the plate is given by:

$$U = \frac{1}{2}\int\int_{\Omega} d(\nabla^2 w)^2 - D(1-\nu)\diamond^4(w,w)dxdy. \qquad (3.29)$$

Following our initial assumptions (i) - (v), the total energy is given by the sum of strain energy, kinetic energy and the potential energy due to the effects of the boundary forces, where

$$U_B = \int_{\partial\Omega}(Q_n w - M_{nn}\frac{\partial w}{\partial n} - M_{ns}\frac{\partial w}{\partial s})ds\ , \qquad (3.30)$$

will denote the potential energy due to the constraint forces applied to the boundary $\partial\Omega$. As before q will denote the external forces (controls) applied only in the interior of Ω. (As usual by "forces" we mean generalized forces, which could have the physical dimension of a force, or of a moment). We do not consider the possibility of an elastic boundary support, in which case a further term would have to be added under the integral sign in (3.30). The possibility of boundary controls may be considered in a later development of this work.

The total energy is given by

$$\mathcal{E}(t) = \frac{1}{2} \int\int_{\Omega} D(\nabla^2 w)^2 - D(1-\nu) \diamond^4 (w,w)dxdy$$

$$+ \int_{\partial\Omega} (Q_n w - M_{nn} \frac{\partial w}{\partial n} - M_{ns} \frac{\partial w}{\partial d})ds + \frac{1}{2}\int\int \rho (\frac{\partial w}{\partial t})^2 dxdy. \quad (3.31)$$

Differentiating both sides of (3.31) with respect to the time variable, we obtain:

$$\frac{d\mathcal{E}(t)}{dt} = \int\int_{\Omega} \{D[(\nabla^2 w)(\nabla^2 v) - (1-\nu) \diamond^4 (w,v)] + \rho v \frac{\partial^2 w}{\partial t^2}\} dxdy$$

$$+ \frac{\partial}{\partial t} \int_{\partial\Omega} (Q_n w - M_{nn} \frac{\partial w}{\partial n} - M_{ns} \frac{\partial w}{\partial s} ds \quad (3.32)$$

(as before: $v = \frac{\partial w(x,y,t)}{\partial t}$) .

With regard to the last term of the formula (3.32), we note that it vanishes identically in the case of a free edge, since then $Q_n = M_{nn} = M_{ns} \equiv 0$ on $\partial\Omega$, and that it also vanishes in the case of a simply supported edge, since then $w = \frac{\partial w}{\partial s} \equiv 0$, and $M_{nn} \equiv 0$ on $\partial\Omega$. In the only remaining case, i.e. the case of a fixed edge, we have $w = \frac{\partial w}{\partial s} \equiv 0$ and $\frac{\partial w}{\partial n} \equiv 0$, and again the entire expression is identically equal to zero on the boundary .

We rewrite the equation (3.32) accordingly:

$$\frac{d\mathcal{E}}{dt} = \int\int \{D[(\nabla^2 w)(\nabla^2 v) - (1-\nu) \diamond^4 (w,v)] + \rho v \frac{\partial^2 w}{\partial t^2}\} dxdv \quad (3.32^a)$$

In the entire discussion, which follows, we shall consider only the cases when the boundary conditions are of the (3.B1), (3.B2), or (3.B3) type, and therefore when $Q_n w = M_{nn} \frac{\partial w}{\partial n} = M_{ns} \frac{\partial w}{\partial s} \equiv 0$ on $\partial\Omega$. We manipulate the formulas (3.31), (3.32) by using the Green's identity:

$$\iint_{\Omega} D(\nabla^2 w)(\nabla^2 w)dxdy = \iint_{\Omega} \nabla^2 (D\nabla^2 w)dxdy + \int_{\partial\Omega} \{D\nabla^2 w \frac{\partial w}{\partial n}\}ds$$

$$- \int_{\partial\Omega} \{w \frac{\partial}{\partial n} (D\nabla^2 w)\}ds.$$

We recall also the formulas for the moments and shears:

$$M_{nn} = -D(\frac{\partial^2 w}{\partial n^2} + \nu \frac{\partial^2 w}{\partial \tau^2}) , \qquad (3.5^{B1})$$

$$M_{ns} = (1-\nu)D \frac{\partial^2 w}{\partial n \partial \tau} , M_{sn} = -M_{ns} , \qquad (3.5^{B2})$$

$$M_{ss} = -D(\frac{\partial^2 w}{\partial \tau^2} + \nu \frac{\partial^2 w}{\partial n^2}) , \qquad (3.5^{B3})$$

$$Q_n = \frac{\partial M_{nn}}{\partial n} + \frac{\partial M_{ns}}{\partial \tau} = - \frac{\partial}{\partial n} [D(\frac{\partial^2 w}{\partial n^2} + \nu \frac{\partial^2 w}{\partial \tau^2})] \qquad (3.7^{B})$$

$$+ \frac{\partial}{\partial \tau} [D(\frac{\partial^2 w}{\partial n \partial \tau} - \nu \frac{\partial^2 w}{\partial n \partial \tau})],$$

$$M_{nn} + M_{ss} = -D(1+\nu)(\frac{\partial^2 w}{\partial n^2} + \frac{\partial^2 w}{\partial \tau^2}) = -D(1+\nu)\nabla^2 w.$$

Hence: $D\nabla^2 w = -\frac{1}{1+\nu}(M_{nn} + M_{ss})$.

We shall need later the following formula:

$$\iint_\Omega D(\nabla^2 w)(\nabla^2 w)dxdy = \iint_\Omega w\nabla^2(D\nabla^2 w)dxdy - \frac{1}{1+\nu}\int_{\partial\Omega}\{(M_{nn}+M_{ss})\frac{\partial w}{\partial n}\}ds$$

$$+ \frac{1}{1+\nu}\int_{\partial\Omega}\{w\frac{\partial}{\partial n}(M_{nn}+M_{ss})\}ds.$$

In the case of the free edge, we have: $M_{nn} \equiv 0$ on $\partial\Omega$.

In the case of a simply supported edge $w \equiv 0$ on $\partial\Omega$, and the second boundary integral vanishes, and finally in the case of a fixed edge, i.e., $\frac{\partial w}{\partial n} \equiv 0$ and $w \equiv 0$ on $\partial\Omega$, and we have:

$$\iint_\Omega D(\nabla^2 w)(\nabla^2 w)dxdy \equiv \iint_\Omega w\nabla^2(D\nabla^2 w)dxdy.$$

Following these rematks we rewrite the formula (3.32) for the general case:

$$\frac{d\mathcal{E}}{dt} = \frac{1}{2}[\frac{\partial}{\partial t}\iint_\Omega \{w\nabla^2(D\nabla^2 w) - D(1-\nu)\diamond^4(w,w)\}\, dxdy]$$

$$- \iint_\Omega (qv+w\frac{\partial q}{\partial t})dxdy + \frac{1}{2}\iint_\Omega v\rho\,\frac{\partial^2 w}{\partial t^2}\,dxdy$$

$$- \frac{1}{1+\nu}\int_{\partial\Omega}(\chi\frac{\partial w}{\partial n} + w\frac{\partial\chi}{\partial n})ds \quad , \qquad (3.33^a)$$

where χ denotes the invariant quantity: $M_{yy} + M_{xx}$. (Note: The proof that $M_{xx} + M_{yy} = M_{nn} + M_{ss}$ for any orthogonal coordinates following the directions n,s comes directly from the well-known fact that the Laplacian $\nabla^2 w$ is invariant under any orthogonal transformation of coordinates.) We use the fact that w(x,y,t) obeys the equation

$$v(\rho \frac{\partial^2 w}{\partial t^2}) = v[q - \nabla^2(D\nabla^2 w) + (1-\nu) \Diamond^4(D,v)], \text{ and obtain:}$$

$$\frac{d\mathcal{E}}{dt} = \frac{1}{2} \iint_\Omega w\nabla^2(D\nabla^2 v) - 2D(1-\nu) \Diamond^4(w,v)dxdy - \iint_\Omega (qv)dxdy$$

$$+ \frac{(1-\nu)}{2} \iint \Diamond^4(D,v)dxdy - \frac{1}{1+\nu} \frac{\partial}{\partial t} \int_{\partial\Omega} (\chi \frac{\partial w}{\partial n} + w \frac{\partial \chi}{\partial n})ds.$$

$$(v = \frac{\partial w}{\partial t}) \qquad (3.33^b)$$

We shall be interested in various modes of optimal control or of excitation, and one of our main problems is to find q, such that either $-\frac{d\mathcal{E}}{dt}$ is maximized in some interval [0,1], that $\mathcal{E}$ assumes some value in the shortest possible time, or that $\mathcal{E}$ assumes an extreme value at some given time $t = \tau$. In all cases the formula (3.33^a) is crucial.

We introduce the following inner product of two admissible transverse displacement functions:

$$\langle w_1, w_2 \rangle = \frac{1}{2} \iint_\Omega [D\nabla^2 w_1 \nabla^2 w_2 - (1-\nu)D \Diamond^4(w_1,w_2)]dxdy$$

$$+ \frac{1}{2} \iint_{\Omega} \rho \frac{\partial w_1}{\partial t} \frac{\partial w_2}{\partial t} \, dxdy \, . \tag{3.34}$$

This product will be used in the later discussion. We see that

$$\langle w_1, w_1 \rangle = U_1(t) + T_1(t) = \mathcal{E}_1(t) \tag{3.34^a}$$

and

$$\frac{d}{dt} \langle w, w \rangle = \frac{d\mathcal{E}}{dt} \, .$$

We need to check the fact that $\langle w_1, w_2 \rangle$ does indeed satisfy all axiomatic requirements of an inner product. That is a routine exercise, and it will be omitted. From the fact that $\langle w_1, w_2 \rangle$ is an inner product, follows immediately the Cauchy-Schwartz inequality:

$$\langle w_1, w_2 \rangle^2 \leq \langle w_1, w_1 \rangle \cdot \langle w_2, w_2 \rangle = \mathcal{E}_1 \mathcal{E}_2$$

which is valid for all $t \in [0,T]$. We note that the exact equality $\langle w_1, w_2 \rangle^2 = \mathcal{E}_1 \mathcal{E}_2$ is true only if there exists a constant C, such that $\nabla^2 w_1 = C\nabla^2 w_2$, $\frac{\partial w_1}{\partial t} = C \frac{\partial w_2}{\partial t}$, and $\diamond^4 (w_1, w_1) = C \diamond^4 (w_2, w_2)$, valid for all $t \in [0,T]$. We compute the time rate change of this product:

$$\frac{d}{dt}\langle w_1,w_2\rangle = \frac{1}{2}\iint_\Omega \{D[\nabla^2 w_1\nabla^2 v_2 + \nabla^2 w_2\nabla^2 v_1]$$

$$- D(1-\nu)[\diamond^4(v_1,w_2) + \diamond^4(v_2,w_1)]$$

$$+ \rho[v_1 \frac{\partial^2 w_2}{\partial t^2} + v_2 \frac{\partial^2 w_1}{\partial t^2}]\}dxdy .$$

We use Green's identity: $\iint_\Omega D\nabla^2 w_1\nabla^2 v_2 dxdy =$

$$\iint_\Omega v_2(\nabla^2 D\nabla^2 w_1)dxdy + \int_{\partial\Omega}\{D\nabla^2 w_1 \frac{\partial v_2}{\partial n} - v_2\frac{\partial}{\partial n}(D\nabla^2 w_1)\}ds,$$

and obtain:

$$\frac{d}{dt}\langle w_1,w_2\rangle = \frac{1}{2}\iint_\Omega \{v_2[\nabla^2(D\nabla^2 w_1) + \rho\frac{\partial^2 w_1}{\partial t^2} - (1-\nu)\diamond^4(D,w_1)]$$

$$+ v_1[\nabla^2(D\nabla^2 w_2) + \rho\frac{\partial^2 w_2}{\partial t^2} - (1-\nu)\diamond^4(D,w_2)]\}dxdy$$

$$+ \frac{1}{2}\iint_\Omega (1-\nu)[v_2\diamond^4(D,w_1) - D\diamond^4(v_2,w_1)$$

$$+ v_1\diamond^4(D,w_2) - D\diamond^4(v_1,w_2)]dxdy$$

$$+ \frac{1}{2}\int_{\partial\Omega}\{D\nabla^2 w_1\frac{\partial v_2}{\partial n} - v_2\frac{\partial}{\partial n}(D\nabla^2 w_1)$$

$$+ D\nabla^2 w_2\frac{\partial v_1}{\partial n} - v_1\frac{\partial}{\partial n}(D\nabla^2 w_2)\}ds . \quad (3.35^a)$$

In the case when D = constant the above formula reduces to:

$$\frac{d}{dt}\langle w_1,w_2\rangle = \frac{1}{2}\iint_\Omega v_2[\nabla^2(D\nabla^2 w_1) + \rho\frac{\partial^2 w_1}{\partial t^2}]$$

$$+ v_1[\nabla^2(D\nabla^2 w_2) + \rho\frac{\partial^2 w_2}{\partial t^2}] - D(1-\nu)[\diamondsuit^4(v_2,w_1)$$

$$+ \diamondsuit^4(v_1,w_2)]\}dxdy + \frac{1}{2}\int_{\partial\Omega}\{D(\nabla^2 w_1)\frac{\partial v_2}{\partial n}$$

$$+ D(\nabla^2 w_2)\frac{\partial v_1}{\partial n} + v_2 Q_{n_1} + v_1 Q_{n_2}\}ds$$

$$= \frac{1}{2}\iint_\Omega (v_2 f_1 + v_1 f_2) - D(1-\nu)\frac{d}{dt}[\diamondsuit^4(w_1,w_2)]\}dxdy$$

$$+ \frac{1}{2}\int_{\partial\Omega}\{D[(\nabla^2 w_1)\frac{\partial v_2}{\partial n} + (\nabla^2 w_2)\frac{\partial v_1}{\partial n}]$$

$$+ v_2 Q_{n_1} + v_1 Q_{n_2}\}\, ds . \qquad (3.35^b)$$

(See the equation 3.7^d).

However, we claim that if D = const. then $\iint D\,\diamondsuit^4(w_1,w_2)dxdy = 0$ for any displacement functions w_1, w_2 which solve the basic equation (3.18). We use the well-known fact that:

$$\iint_\Omega D\,\diamondsuit^4(w,w)dxdy = 0 \quad \text{if} \quad \diamondsuit^4(D,w) = 0.$$

(See for example [9], page 80, equation 6.5, and the discussion of 1.4.)

To prove our claim we first observe that when D = const. we have:

$$\iint_\Omega \diamond^4(w_1,w_2)\,dxdy = \iint_\Omega \{\diamond^4(w_1,w_2) + \tfrac{1}{2}\diamond^4(w_1,w_1) + \tfrac{1}{2}\diamond^4(w_2,w_2)\}dxdy$$

$$= \iint_\Omega \{\frac{\partial^2 w_1}{\partial x^2}\frac{\partial^2 w_2}{\partial y^2} + \frac{\partial^2 w_2}{\partial x^2}\frac{\partial^2 w_1}{\partial y^2} - 2\frac{\partial^2 w_1}{\partial x\partial y}\frac{\partial^2 w_2}{\partial x\partial y}$$

$$+ \frac{\partial^2 w_1}{\partial x^2}\frac{\partial^2 w_1}{\partial y^2} - (\frac{\partial^2 w_1}{\partial x\partial y})^2 + \frac{\partial^2 w_2}{\partial x^2}\frac{\partial^2 w_2}{\partial y^2}$$

$$- (\frac{\partial^2 w_2}{\partial x\partial y})^2\,dxdy$$

$$= \iint_\Omega \{(\frac{\partial^2 w_1}{\partial x^2} + \frac{\partial^2 w_2}{\partial x^2})(\frac{\partial^2 w_1}{\partial y^2} + \frac{\partial^2 w_2}{\partial y^2})$$

$$- [\frac{\partial^2 (w_1+w_2)}{\partial x\partial y}]^2\}dxdy$$

$$= \iint_\Omega \det \begin{bmatrix} \dfrac{\partial^2 (w_1+w_2)}{\partial x^2} & \dfrac{\partial^2 (w_1+w_2)}{\partial x\partial y} \\ \dfrac{\partial^2 (w_1+w_2)}{\partial x\partial y} & \dfrac{\partial^2 (w_1+w_2)}{\partial y^2} \end{bmatrix} dxdy$$

Hence, if we denote by w_3 the displacement $w_3 = \frac{w_1+w_2}{2}$, we can easily obtain

$$\iint_\Omega \det\,(A(w_1+w_2))dxdy = \iint_\Omega \det(A(w_3+w_3))dxdy$$

$$= \iint \diamond^4(w_3,w_3)dxdy$$

$$= 0,$$

and consequently:

$$\iint_{\Omega} \diamond^4 (w_1,w_2)\,dxdy = 0$$

for any w_1,w_2 which solve the equation (3.A1).

The matrix operator A in the above manipulations stood for:

$$A = \begin{bmatrix} \frac{\partial^2}{\partial x^2} , & \frac{\partial^2}{\partial x \partial y} \\ \frac{\partial^2}{\partial x \partial y} , & \frac{\partial^2}{\partial y^2} \end{bmatrix} .$$

This result enables us to simplify our formula (3.35^b) in the case D = const. where upon we obtain:

$$\frac{d}{dt} \langle w_1,w_2 \rangle = \frac{1}{2} \iint_{\Omega} (v_2 f_1 + v_1 f_2)\,dxdy + \frac{1}{2} \int_{\partial\Omega} \{D[(\nabla^2 w_1) \frac{\partial v_2}{\partial n}$$

$$+ (\nabla^2 w_2) \frac{\partial v_1}{\partial n}] + v_2 Q_{n_1} + v_1 Q_{n_2} \} ds. \quad (3.35^c)$$

An identical result is obtained if we assume that $\diamond^4 (D,w) = 0$ for any $w(x,y,t)$ which is a solution of (3.18). Whether much greater generality is obtained is doubtful. It is plausible to conjecture that $\diamond^4 (D,w) = 0$ implies that either D = const. or that D varies linearly in Ω, a case which would imply a variation of the thickness proportional to the cube root of distance.

We can denote by $\{v_1,w_2\}$ the quantity:

$$\{v_1,w_2\} = \frac{1}{2}\iint_\Omega \{D\nabla^2 v_1\nabla^2 w_2 + v_1\nabla^2(D\nabla^2 w_2)$$

$$- v_1(1-\nu)\diamond^4(D,w_2) - D(1-\nu)\diamond^4(v_1,w_2)\}dxdy;$$

then the equality (3.35^a) can be written in the form:

$$\frac{d}{dt}\langle w_1,w_2\rangle = \{v_1,w_2\} + \{v_2,w_1\} + \frac{1}{2}\iint_\Omega (f_2v_1 + f_1v_2)dxdy. \tag{3.35^d}$$

Here f_1, f_2 are the inhomogeneous terms of the respective equations of the form (3.18).

In the particular case when $f_1 \equiv f_2$, $w_1 \equiv w_2$ for all $t \geq 0$,

$$\frac{d\mathscr{E}}{dt} = \frac{d}{dt}\langle w_1,w_1\rangle = 2\{v_1,w_1\} - \iint_\Omega (f_1v_1)dxdy. \tag{3.36}$$

If in addition the plate is freely vibrating, that is: $f_1 \equiv 0$, $w_1 = w_H$, then it follows immediately that the conservation of total energy implies that

$$\{v_H,w_H\} \equiv 0. \tag{3.36^a}$$

Clearly the product $\{v,w\}$ is a function of time only. It is easily proved that the product $\{v,w\}$ is bilinear, and symmetric:

$$\{av,w\} = \{v,aw\} = a\{v,w\} \quad \text{for any constant } a.$$

$$\{(v_1+v_2),w\} = \{v_1,w\} + \{v_2,w\}$$

$$\{v,w_1+w_2\} = \{v,w_1\} + \{v,w_2\}.$$

Hence, using Duhamel's Principle, and representing an arbitrary displacement function $w(x,y,t)$ in the form:

$$w = w_H + q * G,$$

we obtain

$$\{v,w\} = \{(\frac{\partial}{\partial t}(q*G) + \frac{\partial}{\partial t} w_H), (w_H+q*G)\}$$

$$= \{G * \frac{\partial q}{\partial t}, w_H + q * G\} + \{v_H, q * G\}. \qquad (3.37)$$

The displacement function $w_q = q * G$ corresponding to an admissible control q is the solution of the equation $(3.A2^a)$ with the prescribed boundary conditions on $\partial\Omega$ of the form (3.B1) - (3.B2), but with zero initial conditions:

$$w(x,y,0) \equiv 0$$

$\frac{\partial}{\partial t} w(x,y,0) \equiv 0$ in Ω.

(As before * denotes the operation of convolution.)

3.1.10. The Case of $\diamond^4(D,w) \equiv 0$

The above expressions and formula can be greatly simplified if $\diamond^4(D,w) \equiv 0$, which is true in the physically important cases when D = constant in Ω, or when D depends linearly on x and y. The second case occurs in the optimum weight design of plates. If we ignore the effects of the middle forces, the equation (3.A2) becomes:

$$\nabla^2(D\nabla^2 w) + \rho\frac{\partial^2 w}{\partial t^2} = q \qquad (3.A2^b)$$

and if D = constant, this becomes

$$\nabla^4 w + \frac{\rho}{D}\frac{\partial^2 w}{\partial t^2} = \frac{q}{D} \qquad (3.A2^c)$$

The equation $(3.A2^c)$ shows that the Poisson's ratio can not influence the solution $w(x,y,t)$ with a given $q(t)$, D = const., $\rho(x,y)$. (It does influence $D(x,y)$, since $D = (Eh^3)/(12(1-\nu^2))$. It is clear that in any case the Poisson's ratio ν can not explicitly influence the value of the strain energy, but only through D. (See the discussion of Mansfield [9] page 80) and that ν can only affect the deflections and affects U only through the boundary conditions. If $U_B \equiv 0$, then the term $\iint_\Omega D \diamond^4 (w,w)dxdy$ must also be identically equal to zero, and the expression for the strain energy becomes:

$$U = \frac{1}{2} \iint_\Omega D(\nabla^2 w)^2 dxdy. \qquad (3.29^a)$$

(A variational argument for this statement also follows easily. See for example [9], pp. 79-82, or [12].)

A similar conclusion is reached in the case when $\diamond^4 (D,w) \equiv 0$ even if $D \not\equiv$ constant in Ω. The product $\langle w_1,w_2\rangle$ assumes the form:

$$\langle w_1,w_2\rangle = \frac{1}{2} \iint_\Omega D(\nabla^2 w_1)(\nabla^2 w_2)dxdy$$

$$+ \frac{1}{2} \iint_\Omega \rho \frac{\partial w_1}{\partial t} \frac{\partial w_2}{\partial t} dxdy . \qquad (3.34^a)$$

The rate of change of this product is given by (3.35^c) below

$$\frac{d}{dt} \langle w_1,w_2\rangle = \iint_\Omega (v_1 f_2 + v_2 f_1)dxdy$$

$$- \frac{1}{2} \int_{\partial\Omega} (v_1 \frac{\partial}{\partial n} (D\nabla^2 w_2) + v_2 \frac{\partial}{\partial n} (D\nabla^2 w_1)$$

$$- (D\nabla^2 w_2)\frac{\partial v_1}{\partial n} - (D\nabla^2 w_1)\frac{\partial v_2}{\partial n})ds.$$

Using the formula (3.5^a) and (3.5^c), we substitute:

$$\chi_1 = M_{xx_1} + M_{yy_1} = -D(1+\nu)\nabla^2 w_1 ,$$

and

$$\chi_2 = -D(1+\nu)\nabla^2 w_2 ,$$

to obtain:

$$\frac{d}{dt}\langle w_1,w_2\rangle = \frac{1}{2}\iint_\Omega (v_1 f_2 + v_2 f_1)dxdy$$
$$- \frac{1}{2(1+\nu)}\int_{\partial\Omega} (\chi_2 \frac{\partial v_1}{\partial n} + \chi_1 \frac{\partial v_2}{\partial n} - v_1 \frac{\partial \chi_2}{\partial n}$$
$$- v_2 \frac{\partial \chi_1}{\partial n})ds. \qquad (3.35^c)$$

Recalling the relationship (3.7^d) if D = const. we can also rewrite (3.35^c) in the form:

$$\frac{d}{dt}\langle w_1,w_2\rangle = \frac{1}{2}\int_{\partial\Omega} (v_1 f_2 + v_2 f_1)dxdy$$
$$- \frac{1}{2}\int_{\partial\Omega} [v_1 Q_{n_2} + v_2 Q_{n_1} - \frac{1}{1+\nu}(\chi_1 \frac{\partial v_2}{\partial n}$$
$$+ \chi_2 \frac{\partial v_1}{\partial n})]ds \qquad (3.35^d)$$

where Q_{n_i} are the shear forces which are related to the moments by the equations (3.7^a), (3.7^b), and (3.7^c).

We note that in the case of clamped edge (condition 3.B1)) the equation (3.35^d) reduces to:

$$\frac{d}{dt}\langle w_1,w_2\rangle = \frac{1}{2}\iint_{\Omega} (v_1f_2 + v_2f_1)dxdy, \qquad (3.35^e)$$

since in this case $v_1 = v_2 = \frac{\partial v_1}{\partial n} = \frac{\partial v_2}{\partial n} \equiv 0$ on $\partial\Omega$, and the contour integral vanishes.

The next result will be used in proving the basic Theorem (3.2.2). For this reason we shall state it as a lemma.

Lemma 3.1.1.

Let $f(x,y,t)$ be an admissible control and $w(x,y,t)$ be the corresponding deflection of a plate, whose flexural rigidity D and density ρ are constant. Let w_H represent the solution of the homogeneous equation $(3.A1^a)$. Let both $w(x,y,t)$ and $w_H(x,y,t)$ satisfy the condition $w = 0$ on $\partial\Omega$ and $w_H = 0$ on $\partial\Omega$ (the boundary of the plate). Then

$$\frac{d}{dt}\langle w,w_H\rangle = \frac{1}{2}\int_{\Omega} (w_Hf)dxdy + \frac{D}{2}\int_{\partial\Omega} \nabla^2w\frac{\partial v_H}{\partial n} + \nabla^2w_H\frac{\partial v}{\partial n})ds,$$

where as before:

$$v = \frac{\partial w}{\partial t} \;, \quad v_H = \frac{\partial w_H}{\partial t} \;.$$

The proof follows from the formula (3.35^b) upon substituting: $f_1 = f$, $f_2 \equiv 0$, $w_1 = w$, $w_2 = w_H$ and from the observation that

$$\iint_\Omega \diamondsuit^4 (w, w_H) dxdy = 0.$$

3.2 Optimal Control Principles for the Small Deflection Theory in the Mixed Boundary and Initial Value Problems for Thin Plates.

3.2.1. Statement of the Control Problems

a) The minimal time control problem

We consider the equations (3.A1), (3.A2) and the corresponding MBVP. The initial conditions (3.C1) and (3.C2) determine the initial value of the total energy $\mathcal{E}_0 = \mathcal{E}(0)$. Given a real number $0 \le \hat{E} < \mathcal{E}_0$ find an admissible control $q(x,y,t)$ such that the total energy of the vibrating plate $\mathcal{E}(t)$ is reduced to the value $\hat{E}$ in the shortest possible time.

b) A similar problem is defined below as the minimal time excitation problem

Again the initial conditions determine the initial value of the total energy $\mathcal{E}_0$. We allow $\mathcal{E}_0 \ge 0$. Given a real number $\hat{E} > \mathcal{E}_0$ find admissible control $q(x,y,t)$ such that the total energy of the vibrating plate $\mathcal{E}(t)$ is raised to the value $\hat{E}$ in the shortest possible time.

c) The fixed time interval optimal control (or simply optimal interval control.)

Given suitable boundary and initial conditions of the MBVP, and given a time interval [0,T] find an admissible control $\tilde{q}(x,y,t)$, such that the total energy of the plate is reduced (raised) to the lowest (highest) possible level at the time $t = T$, i.e. $(\mathcal{E}(\tilde{q}(x,y,t),T) \leq \mathcal{E}(q(x,y,t),T)$ for any admissible control $q(x,y,t)$, (or in the excitation problem $\mathcal{E}(\tilde{q}(x,y,t),T) \geq \mathcal{E}(q(x,y,t),T)$ for any admissible control $q(x,y,t)$.

Remark

Optimal controls are generally not unique. However, some form is non-uniqueness turns out to be acceptable, but other forms will defeat the whole idea of a meaningful control. Consider the case of an optimal interval control problem, where the interval [0,T] is sufficiently long, and we find that not only there exists a control $\tilde{q}_1(x,y,t)$ such that $\mathcal{E}(q_1(x,y,t),T) = 0$, but that it is possible to let the plate vibrate freely for the length of time t_1, and then apply an admissible control $\tilde{q}_2(x,y,t)$ which will reduce the total energy to the zero level at the time T. (See [17] for an analogous discussion).

This situation is illustrated on Figure 1 of Chapter 1.

Our optimal control theorems, and in particular the Pontryagin's principle will turn out to be devoid of any information. We shall attempt to avoid the situation illustrated on Figure 1 , Chapter 1, by insisting that in the optimal control problem for the interval [0,T] the total energy can not be reduced to the zero level at the time T by any admissible control.

The theorem 4 of Chapter 2 illustrates the fact that the minimal time and the fixed interval control problems are closely related. We shall state it again for the two-dimensional case.

Lemma 3.1.

We consider the MBVP with suitable initial and boundary conditions. Let $\dot{q}(x,y,t)$ be an optimal time control reducing the total energy from the initial value $\mathcal{E}(0)$ to a given value $\hat{E}$ $0 < \hat{E} < \mathcal{E}(0)$ in the shortest possible time $T > 0$. Then $\check{q}(x,y,t)$ is also an optimal interval control for the fixed interval [0,T].

Remarks

The converse of this lemma is clearly false, as can be seen from the example illustrated on Figure 1, Chapter 1, it is also clear that any time optimal excitation is a fixed time interval optimal excitation. The proof is analogous to the proof of Theorem 4, Chapter 1 with the inequalities appropriately reversed.

The above lemma also implies that it will suffice to develop the control principles for the fixed interval case, since any optimal time control will also be a fixed interval optimal control and the validity of our results will be preserved.

3.2.2. Some Mathematical Preliminaries

In the problems of existence of an optimal control we shall find very useful the lemma 2 given below, which is a simplification of the lemma on completeness of the space K* of generalized functions as given by Gelfand and Shilov in Appendix A, pages 368-369 of [5] volume 1. For the sake of completeness, we shall define below the concept of convergence in the space K*, and state the lemma of Gelfand and Shilov. Let Ω be any bounded region of E^2 (or of E^n).
The set (vector space) K of test functions on Ω is the set of all real valued functions $\phi(x,y)$ which possess derivatives of all orders in Ω and are identically equal to zero outside of Ω.
(The example of such a function is given by

$$\phi(x,y) = \exp\left(-ar/(a^2-r^2)\right) \quad \text{for } r < a$$

$$\phi(x,y) \equiv 0 \quad \text{for } r \geq a$$

where $r = \sqrt{x^2+y^2}$, and where Ω is assumed to contain the origin, while the number a is chosen smaller than the distance from the origin to $\partial\Omega$.) The space K^* of generalized functions over the space of test functions K is the set of all continuous linear functionals defined on K. A sequence of generalized functions $\{f_i\} \in K^*$ is said to converge to $f \in K^*$ if for every test function $\phi \in K$ $\lim_{i\to\infty} (f_i,\phi) = (f,\phi)$.

The lemma of Gelfand-Shilov states that the space K^* is complete, i.e. given a sequence of generalized functions $f_1, f_2, \ldots, f_n, \ldots$, such that $\lim_{i\to\infty} (f_i,\phi)$ exists for every $\phi \in K$, then $f(\phi) = \lim_{i\to\infty} (f_i,\phi)$ defines again a continuous linear functional f on K.

In this work we shall need a somewhat similar result:

Lemma 3.2.

Let R denote the class of functions $u(x,y,t)$, and of the time derivatives $\frac{\partial u}{\partial t}$ of such functions obeying the "elastica" hypothesis in Ω, and satisfying the condition that either the displacement $w(x,y,t)$ or the moment $M_{nn}(w(x,y,t))$ vanishes on $\partial\Omega$ for all $t \in [0,\infty)$, where M_{nn} is defined by the equation (3.5^{B1}).

Let R^* the space of all continuous linear functionals mapping elements of R into the real line. We shall consider only a subset $\hat{R}^* \subset R^*$ of all such functionals f obeying $||f||_\Omega \leq 1$, where $||f||_\Omega = \sup_{||u||=1} |(f,u)|$. (If

$\sup_{||u||=1} |(f,u)_\Omega|$ does not exist we assign $||f||_\Omega = \infty$.)

Then we assert that the space R* is complete. The proof of this lemma is more elementary than of the lemma of Gelfand-Shilov stated above, but follows an identical line of argument. (See [5], Appendix A, pages 368-369).

3.2.3. The Basic Convexity Lemma

Lemma 3.2.3a

The set of admissible controls is convex, (i.e. if f_1, f_2 are admissible controls then $\lambda f_1 + (1-\lambda)f_2$ is also an admissible control for any $0 \leq \lambda \leq 1$.)

Corollary

Combining the result of this lemma with the Duhamel's principle (3.28) we obtain an important corollary: The set of all admissible transverse displacements is convex. (i.e. if $w_1(x,y,t)$ is an admissible transverse displacement corresponding to an admissible control f_1, and if $w_2(x,y,t)$ is an admissible displacement corresponding to an admissible control f_2, then $w = \lambda w_1 + (1-\lambda)w_2$ is also an admissible displacement for any $0 \leq \lambda \leq 1$. Of course, w_1, w_2 are assumed to be solutions obeying the stated boundary and initial conditions of MBVP).

3.2.4. The Uniqueness of the Finite State

In general the optimal controls (corresponding to the various definitions of optimality) are non-unique, and the corresponding transverse displacements are non-unique. In spite of this, certain weaker forms of uniqueness theorems hold for the transverse displacements which correspond to optimal controls.

Lemma 3.2.4. (uniqueness of the finite state)

Let us assume that no energy is transmitted at the boundary. Let $f_1(x,y,t)$, $f_2(x,y,t)$ be two admissible controls, which are optimal controls for the $[0,\tau]$ fixed time interval. Then the corresponding shapes of the plate, and the corresponding velocities coincide at the time $t = \tau$, i.e.

$$w_1(f_1,x,y,\tau) = w_2(f_2,x,y,\tau)$$

and

$$\frac{\partial w_1(f_1,x,y,\tau)}{\partial t} = \frac{\partial w_2(f_2,x,y,\tau)}{\partial t} \tag{3.36}$$

Proof: Let us assume to the contrary that $\underset{\sim}{w}_1 \neq \underset{\sim}{w}_2$ at the time

where $\underset{\sim}{w} = \begin{cases} w(x,y,t) \\ \frac{\partial w}{\partial t}(x,y,t) \end{cases}$. Let us denote by $\tilde{\mathcal{E}}(\tau)$ the lowest

value of total energy attainable at the time τ. By the convexity of the set of admissible displacements we conclude that $w(x,y,t) = \frac{1}{2}(w_1+w_2)$ is also an admissible displacement.

The corresponding total energy at the time $t = \tau$ is given by the equation (3.29).

$$\mathcal{E}(w,\tau) = U(\tau) + T(\tau) = \frac{1}{2}\iint_\Omega \{\frac{1}{4} D(\nabla^2 w_1+\nabla^2 w_2)^2$$

$$- \frac{1}{4} D(1-\nu) \diamondsuit^4 (w_1+w_2,w_1+w_2)\}dxdy$$

$$+ \frac{1}{2}\iint \frac{1}{4} \rho(\frac{\partial(w_1+w_2)}{\partial t})^2 dxdy$$

$$= \frac{1}{8}\{\iint_\Omega [D(\nabla^2 w_1)^2 + D(\nabla^2 w_2)^2 - D(1-\nu) \diamondsuit^4 (w_1,w_1)$$

$$- D(1-\nu) \diamondsuit^4 (w_2,w_2) + \rho(\frac{\partial w_1}{\partial t})^2 + \rho(\frac{\partial w_2}{\partial t})^2]dxdy$$

$$+ 2\iint_\Omega [D(\nabla^2 w_1)(\nabla^2 w_2) - D(1-\nu) \diamondsuit^4 (w_1,w_2)$$

$$+ \rho\frac{\partial w_1}{\partial t}\frac{\partial w_2}{\partial t}] dxdy\}$$

$$= \frac{1}{8}\{4\tilde{\mathcal{E}}(\tau) + 4\langle w_1,w_2\rangle_{t=\tau} = \frac{1}{2}\tilde{\mathcal{E}}(\tau) + \frac{1}{2}\langle w_1,w_2\rangle_{t=\tau} .$$

By the Cauchy-Schwartz inequality $\langle w_1,w_2\rangle \leq \sqrt{\langle w_1,w_1\rangle \langle w_2,w_2\rangle}$. Since $\langle w_1,w_1\rangle_{t=\tau} = \langle w_2,w_2\rangle_{t=\tau} = \tilde{\mathcal{E}}(\tau)$ we obtain:

$$\mathcal{E}(w,\tau) = \frac{1}{2}\tilde{\mathcal{E}}(\tau) + \frac{1}{2}\langle w_1,w_2\rangle_{t=\tau} \le \frac{1}{2}\tilde{\mathcal{E}}(\tau) + \frac{1}{2}\tilde{\mathcal{E}}(\tau)$$

$$= \tilde{\mathcal{E}}(\tau).$$

However, $\tilde{\mathcal{E}}(\tau)$ was the lowest possible level of total energy attainable at the time $t = \tau$. Hence, we must have the strict equality:

$$\mathcal{E}(w,\tau) = \tilde{\mathcal{E}}(\tau).$$

This equality means that

$$\langle w_1,w_2\rangle^2 = \langle w_1,w_1\rangle \langle w_2,w_2\rangle$$

which implies that $w_1 = \alpha w_2$ where α is some constant; but the only suitable constant turns out to be $\alpha = 1$, and the uniqueness of the final condition is established.

3.2.4a. Discussion of the Uniqueness of the Optimal Excitation

We observe that the above arguments fail in the case of the optimum excitation of a plate. Again let us denote by $\tilde{\mathcal{E}}(\tau)$ the greatest level of energy attainable at the time $t = \tau$ following an admissible excitation of the plate. Let $\tilde{f}(x,y,t)$ be an optimal excitation (assuming that it exists) and $\tilde{w}$ be the corresponding transverse displacement, $\tilde{w} = \tilde{w}(\tilde{f})$. Then if two such optimal admissible excitation functions exist, say $\tilde{f}_1, \tilde{f}_2$, so that $\mathcal{E}(\tilde{f}_1,\tau) = \mathcal{E}(\tilde{f}_2,\tau) = \tilde{\mathcal{E}}$, then $\Lambda f_1 + (1-\Lambda)f_2$ is again an admissible excitation, with $\underset{\sim}{w} = \Lambda \underset{\sim}{w}_1 + (1-\Lambda)\underset{\sim}{w}_2$ being the corresponding displacement. (See the equation (1.28)).

Then $\mathcal{E}(w,\tau) = \frac{1}{2}\tilde{\mathcal{E}}(\tau) + \frac{1}{2}\langle w_1, w_2 \rangle_{t=\tau} \leq \tilde{\mathcal{E}}(\tau)$, if we substitute $\Lambda = \frac{1}{2}$. And the strict inequality must be true if $\underset{\sim}{w}_1 \neq \underset{\sim}{w}_2$ at the time τ, since then $\langle w_1, w_2 \rangle < \tilde{\mathcal{E}}$. Hence, if f_1 and f_2 are optimal excitations $\frac{1}{2}(f_1+f_2)$ cannot be optimal if $\underset{\sim}{w}_1(\tau) \neq \underset{\sim}{w}_2(\tau)$. In fact, $\Lambda f_1 + (1-\Lambda)f_2$ cannot be optimal if $\underset{\sim}{w}_1(\tau) \neq \underset{\sim}{w}_2(\tau)$ for any $0 < \Lambda < 1$. This lack of convexity of the set of optimal excitations prevents us from following an identical argument, and reaching a conclusion analogous to the optimal control case. For a more complete discussion see the section on optimal excitation of Chapter 2.

3.2.5. Pontryagin's Principle

3.2.5.0. Some Preliminary Remarks

In this section we establish the maximality principles analogous to Chapter 2, and to the so-called Pontryagin's principle as given in [26], for the control of ordinary differential equation, and analogous to the maximum principle developed for the special cases in the theory of partial differential equations by Butkovskii [6], Egorov [12], Russell [32]. The formulation of Pontryagin's principle for thin plates with uniform rigidity and density, as given in §2.5.1 of this monograph in complete analogy with the corresponding formulation for the vibrating beam as given by the author in [17]. The more complex formulation given in §3.2.5.4 and §3.2.5.5 results from the presence of the terms of the form $\diamondsuit^4(\cdot,\cdot)$ which do not occur in the beam theory, and from our inability to integrate by parts on the boundary of the region Ω, and then omit the troublesome boundary terms.

Even more complex formulation of Pontryagin's principle would result if we drop the assumption that no energy is transmitted at the boundary. Since the complexity of the problem increases immensely with the removal of each assumption, for the sake of clarity, we shall first discuss the maximal principle in the simplest possible case, then formulate the increasingly more complex cases, rather than attempting to formulate it in the most general case, and derive the simpler cases by ignoring appropriate terms of the general expression. In all cases discussed below in §2.5.1, §2.5.2, §2.5.3, we shall assume that no energy transfer occurs at the boundary of the plate.

3.2.5.1.

Pontryagin's principle for the case D = constant, ρ = constant with boundary conditions of the type (B1). The total strain energy of the plate is then given by:

$$U = \frac{1}{2} \iint_{\Omega} D(\nabla^2 w)^2 dxdy = \frac{D}{2} \iint_{\Omega} (\nabla^2 w)^2 dxdy \quad ,$$

and the kinetic energy by

$$T = \frac{1}{2} \rho \iint (\frac{\partial w}{\partial t})^2 dxdy.$$

The total energy is $\mathcal{E} = U + T$. The total rate of change of the total energy is given by the formula (3.32^a) with the term $\diamond^4 (w,v)$ vanishing. Since the energy is conserved, we obtain:

$$0 = \frac{d\mathcal{E}}{dt} = D \iint_{\Omega} (\nabla^2 w)(\nabla^2 v) dxdy$$

$$+ \rho \iint_{\Omega} (v \frac{\partial v}{\partial t}) dxdy. \qquad (3.2.1)$$

The formula (3.35^c) is applicable in this case:

$$\frac{d}{dt} \langle w_1, w_2 \rangle = - \frac{1}{2(1+\nu)} \int_{\partial\Omega} \{ \chi_2 \frac{\partial v_1}{\partial n} + \chi_1 \frac{\partial v_2}{\partial n} + v_1 \frac{\partial \chi_2}{\partial n}$$

$$+ v_2 \frac{\partial \chi_1}{\partial n} \} ds + \frac{1}{2} \iint_{\Omega} (v_1 f_2 + v_2 f_1) dxdy.$$

It is clear that in the case when condition (B1) is valid on $\partial\Omega$ (a fixed edge) the equation (3.35^d) is valid ($v \equiv 0 \equiv \frac{\partial v}{\partial n}$ on $\partial\Omega$), and

$$\frac{d}{dt} \langle w_1, w_2 \rangle = \frac{1}{2} \iint (v_1 f_2 + v_2 f_1) dxdy \; . \qquad (3.2.2)$$

In particular if $f_2 \equiv 0$, that is if $w_H = w_2$ is a solution of the homogeneous equation ($A3^a$), we have:

$$\langle w_1, w_H \rangle = \frac{1}{2} \iint_\Omega (v_H f_1) dxdy. \qquad (3.2.2^a)$$

We are now prepared to state the simplest version of Pontryagin's principle for thin plates:

<u>Theorem 3.2.1.</u>

Let us assume that $\tilde{\phi}(x,y,t)$ is an optimal control on the fixed time interval $[0,T]$ for a thin homogeneous plate, whose flexural rigidity and density are constant in the domain Ω of the plate. Let the plate's edge be clamped along the entire $\partial\Omega$. (That is on $\partial\Omega$ the displacement function $w(x,y,t)$ satisfies the condition (B1):

$$w \equiv 0, \; \frac{\partial w}{\partial n} \equiv 0.)$$

Let $w_H(x,y,t)$ denote the displacement of this plate vibrating freely, so that the final conditions at the time T are:

$$w_H(x,y,T) = \tilde{w}(\tilde{\phi}(x,y,t),T)$$

$$\frac{\partial w_H(x,y,T)}{\partial t} = \frac{\partial \tilde{w}(\tilde{\phi}(x,y,t),T)}{\partial t}$$

Then the following inequality is true:

$$\iint_{\Omega} [-\tilde{\phi}(x,y,t)\frac{\partial w_H(x,y,t)}{\partial t}]dxdy \geq \iint_{\Omega} [-f(x,y,t)\frac{\partial w_H(x,y,t)}{\partial t}]dxdy \tag{3.2.3}$$

for all $t \in [0,T]$, where $f(x,y,t)$ is any admissible control.

We note that this statement is completely analogous to the Theorem 3 of [17]. The proof turns out to be a repetition of the proof given in [17] and for that reason shall be omitted. As in [17] the equation (2.2) is crucial in the proof of (2.3). A detailed proof of the more complex case will be given in the Theorem (2.3). Since this theorem is a special case of the Theorem 2.3, we shall omit the proof at this stage.

Let us now observe that the Pontryagin's principle as given by the inequality (2.3) is inapplicable, if $\mathcal{E}(\hat{\phi}(x,y,t),T) = 0$. If the total energy of the plate can be reduced to zero at the time T, then $\tilde{w}_H(x,y,t) \equiv 0$, $t \in [0,T]$, and clearly the inequality (2.3) is meaningless. However, if $\mathcal{E}(\hat{\phi}(x,y,t),T) = 0$ but $\mathcal{E}(\hat{\phi}(x,y,t),\tau) > 0$ for any $0 < \tau < T$ it is possible to introduce a sequence of optimal controls $\{\phi_i\}$ converging to $\hat{\phi}(x,y,t)$ with the inequality (3.2.3) applicable to each element $\hat{\phi}_i$

of that sequence. A detailed description of this limiting process will not be given here.

We observe also the usual shortcomings of Pontryagin's principle. To affect a comparison of an arbitrary control with supposedly an optimal control we need to know the final state of the vibrating plate obtained after the application of an optimal control. Again, however, this principle may be useful in a negative way. That is we can use the inequality (2.3) to demonstrate that some control $\phi(x,y,t)$ is <u>not</u> an optimal control.

Example

Let us consider a homogeneous circular plate subjected to a uniformly distributed load of intensity p_0. The edge is clamped. At the time $t = 0$ the load is suddenly removed. The initial deflection is then given by:

$$w(r,0) = \frac{p_0 R^4}{64\pi D}\left[1 - \frac{r^2}{R^2}\right]^2$$

$$(r = \sqrt{x^2+y^2}\ , \ r \leq R).$$

It is clear that $w(0,0) =$ maximum $w(r,0)$, $0 \leq r \leq R$.

A control consisting of a constant load $\hat{\phi}(x,y,t) = Cp_0$ is suggested for the fixed time interval $[0,T]$, $T = \frac{1}{4}\, n_1$ with the constant C chosen to be $C = \dfrac{1}{\pi p_0 R^2}$, to assure: $\iint_\Omega |\hat{\phi}|\,dxdy = 1$.

The time interval n_1 selected above corresponds to one free vibration cycle of the plate. We observe that the freely vibrating plate will vibrate with the angular velocity $\omega = \frac{4T}{2\pi}$, we also note that the average velocity will be distributed in the same manner as $w(r,0)$ and that

$$w_H(r,T) = \frac{R^2}{64\pi D}\left[1 - \frac{r^2}{R^2}\right]^2$$

and

$$v_H(r,T) = \frac{\partial w_H(r,T)}{\partial t} = \omega[w(r,0)-w_H(r,T)].$$

The only important detail emerging from these formulas is the intuitively obvious fact that

$$\frac{\partial w_H(0,t)}{\partial t} > \frac{\partial w_H(r,t)}{\partial t}, \quad r \neq 0 \quad t \in [0,T].$$

Hence, the sign of our control load was correct, but its distribution certainly was not optimal. Choosing for example a time independent admissible load: $f_1(x,y,t) = 4Cp_0$ when $0 \leq r \leq \frac{R}{2}$, and $f_1(x,y,t) = 0$ when $\frac{R}{2} \leq r \leq R$ (C is given as before by: $C = 1/(\pi p_0 R^2)$, we obtain:

$$\iint_\Omega - f_1(x,y,t)\frac{\partial w_H(x,y,t)}{\partial t}\,dxdy > \iint_\Omega \left(-\hat{\phi}\,\frac{\partial w_H}{\partial t}\right)dxdy,$$

showing that $\hat{\phi}$ was not an optimal control.

A gradual improvement technique using the standard form of Pontryagin's principle was discussed by the author in [17] in the case of a vibrating inhomogeneous beam. Clearly, the same technique, which is also suggested by the above example can be used to improve a control in a number of iterative steps. In this example taking f_1 as our candidate for the optimal control, we could compute again $w_H(r,T)$, $v_H(r,T)$ (which would be different this time) and suggest a new improved control $f_2(x,y,t)$, etc..

The technique is very tedious, and in addition we should point out that while such iteration results in improvements of some arbitrarily selected control, we offer no assurance that this iterative process will result in the controls $f_i(x,y,t)$ converging to the optimal control.

The above iteration technique must not be confused with the iteration technique discussed later for the instantly optimal control, where we not only show that controls $f_i(x,y,t)$ do converge, but also prove that the limit is independent of the manner in which the iteration was performed, and also is independent of the choice of the original interval $[0,T]$. We remark that the choice of "optimal" control was made in this example in a deliberately clumsy manner. It could be deduced from a physical argument that initially the optimal control had to be a point load (Dirac delta function) applied at the center of the plate.

3.2.5.2.

Pontryagin's principle for the homogeneous plate (D = const., ρ = const.) with a simply supported part of the boundary consisting of straight lines. (This is a special case of the subsequent theorem (2.3).) Again it can be shown that with D = const. the strain energy of the plate is independent of the Poisson ratio, and is therefore given by

$$U = \frac{1}{2} \iint_{\Omega} D(\nabla^2 w)^2 dxdy.$$

The expression (3.35^c) for the product $\frac{d}{dt} \langle w_1, w_2 \rangle$ becomes:

$$\frac{d}{dt} \langle w_1, w_2 \rangle = \frac{1}{2} \iint_{\Omega} (v_1 f_2 + v_2 f_1) dxdy$$

$$- \frac{1}{2(1+\nu)} \int_{\partial\Omega} (\chi_1 \frac{\partial v_2}{\partial n} + \chi_2 \frac{\partial v_2}{\partial n}) ds.$$

Assuming that $f_2 \equiv 0$ ($w_2 = w_H$), we have:

$$\frac{d}{dt} \langle w_1, w_H \rangle = \frac{1}{2} \iint_{\Omega} (v_H f_1) - \frac{1}{2(1+\nu)} \int_{\partial\Omega} (\chi_1 \frac{\partial v_H}{\partial n} + \chi_H \frac{\partial v_1}{\partial n}) ds. \quad (3.2.4)$$

As we have remarked following the development of equation (3.35[c]) the contour integral in the equation (3.2.4) does not have to vanish if the boundary of the plate is only simply supported. In an exceptional case when a part of the simply supported boundary (say Γ_1) is a straight line and $D \neq 0$ on Γ_1, we have

$$\int_{\Gamma_1} (\chi_1 \frac{\partial v_2}{\partial n} + \chi_2 \frac{\partial v_1}{\partial n})dx = 0 \tag{3.2.5}$$

because $\chi_1 = \chi_2 \equiv 0$ on Γ_1 independently of the controls f_1, f_2. This statement follows quite easily from the observation that $M_{nn} = 0$ on Γ_1 is equivalent to the statement:

$$D(\frac{\partial^2 w}{\partial n^2} + \nu \frac{\partial^2 w}{\partial \tau^2}) = 0. \qquad (*)$$

Since Γ_1 is a straight line, we can replace $\frac{\partial^2 w}{\partial \tau^2}$ by $\frac{\partial^2 w}{\partial s^2}$ on Γ_1. However, since $w \equiv 0$ on $\partial\Omega$ (the plate is assumed to be simply supported), we have $\frac{\partial w}{\partial s} \equiv \frac{\partial^2 w}{\partial s^2} \equiv 0$ on $\partial\Omega$. This implies that $\frac{\partial^2 w}{\partial s^2} = \frac{\partial^2 w}{\partial \tau^2} \equiv 0$ on Γ_1. Now it follows from (*) that $\frac{\partial^2 w}{\partial n^2} \equiv 0$ on Γ_1, since by our previous assumption $D \neq 0$ on Γ_1. Therefore, $\nabla^2 w \equiv 0$ on Γ_1 which in turn implies $\chi = 0$ on Γ_1, as we have claimed. Hence, the integral (3.2.5) must be equal to zero.

It follows easily now that the inequality (3.2.3) is

applicable to the cases when $\partial\Omega$ consists of subarcs Γ_1 and Γ_2 such that $\overline{\Gamma_1 \cup \Gamma_2} = \partial\Omega$, and Γ_1 is the simply supported part of the boundary (condition B2) consisting of a straight line, while Γ_2 is the part of the boundary (not necessarily straight) on which the edge is clamped (condition (B1)).

We intend to show that the inequality (3.2.3) is also applicable to the physically important case when the simply supported part of the boundary meets the clamped part of the boundary at a corner point. Let us now state the following:

Theorem 3.2.2.

Let us assume that the boundary of Ω consists of a finite collection of smooth arcs Γ_1, such that condition (B1) is satisfied on Γ_1 (i.e. $w \equiv 0$, $\frac{\partial w}{\partial n} \equiv 0$ on Γ_1) and of finite number of line segments Γ_2, such that the plate is freely supported on Γ_2 (the condition (B2)). Let us assume that all corners are internal corners, i.e. the angle contained in between Γ_2 and the tangent to Γ_1 at the corner point does not exceed π. Let $\hat{\phi}(x,y,t)$ be an optimal (admissible) control on the fixed time interval $[0,T]$ and let $f(x,y,t)$ be any admissible control. Then the inequality (3.2.3) holds, i.e.

$$\iint_\Omega [-\hat{\phi}(x,y,t) \frac{\partial w_H(x,y,t)}{\partial t}]dxdy$$

$$\geq \iint_\Omega [-f(x,y,t) \frac{\partial w_H(x,y,t)}{\partial t}]dxdy,$$

where $w_H(x,y,t)$ has the same meaning as in the statement of the theorem 2.1.

Proof

Let us replace the corner points by circle segments of radius $\varepsilon_i = \frac{1}{2^i}$ with i chosen sufficiently large to permit such change.

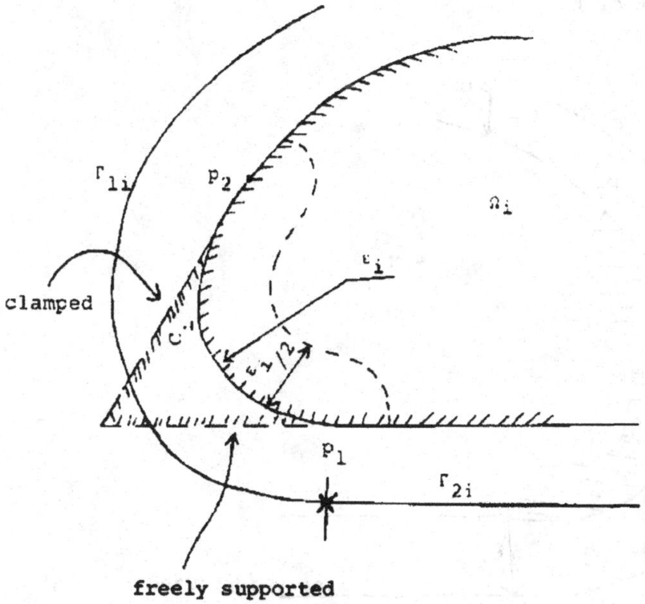

Figure 3[a]

The segment of the circle drawn with the radius ε_i is contained in Ω, and is tangential to the arc of Γ_1 at a point p_2, and to line Γ_2 at a point p_1, as shown on figure 3[a]. The rounded part of the boundary, that is the circular arc p_1p_2 will be denoted by C_i. The modified region now occupied by the plate (with all corners rounded off) will be denoted by Ω_i.

We assume that conditions (B1) will be satisfied on C_i and on the unchanged part of Γ_1, while condition (B2) is satisfied on the unchanged part of Γ_2.

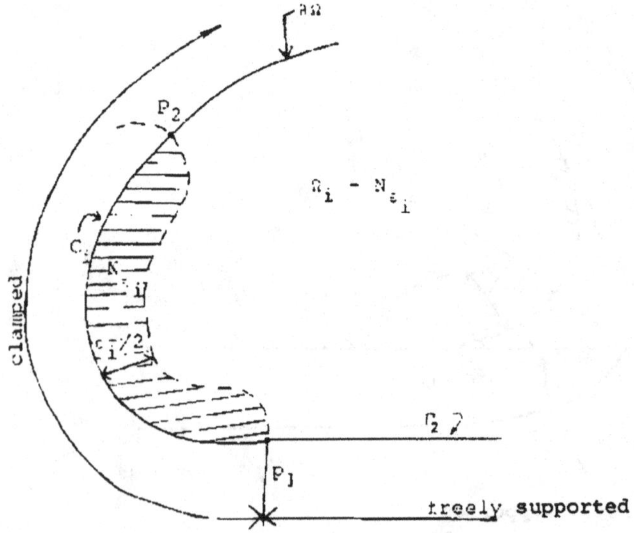

Figure 3[b]

The conditions are now satisfied for the correctness of the inequality

$$\iint_{\Omega_i} [\hat{\phi}_i \frac{\partial w_{H_i}}{\partial t}]dxdy \geq \iint_{\Omega_i} (-f_i \frac{\partial w_{H_i}}{\partial t})dxdy$$

where $\hat{\phi}_i$ is an optimum control for the region Ω_i, f_i is an arbitrary admissible control for Ω_i, w_{H_i} is the solution of the homogeneous equation of MBVP, satisfying the same final condition as $w(\phi_i(x,y,t))$. In the region Ω_i the boundary conditions are posed as stated above and as illustrated on figure 3^b. The initial conditions are as follows: Let N_{ε_i} denote the $\varepsilon_i/2$ neighborhood of the rounded corner. Then in the region $\Omega_i \setminus (\Omega_i \cap N_{\varepsilon_i})$ we have:

$$w(x,y,0) = \psi(x,y)$$

$$\frac{\partial w(x,y,0)}{\partial t} = \eta(x,y),$$

which are the specified initial conditions for $w(x,y,t)$ in Ω as given in the initial conditions (C1), (C2), while in $\Omega_i \cap N_{\varepsilon_i}$ we apply a mollifier function χ of class C^∞) which meets both the conditions C1, C2 on the boundary of N_{ε_i} and the condition

$$\begin{cases} w(x,y,0) = 0 \\ \dfrac{\partial w}{\partial n}(x,y,0) = 0 \end{cases}$$

on C_i. (We recall that C_i is the rounded part of $\partial\Omega$ drawn with the radius ε_i.) (For example a function of the type:

$$\phi(r,\varepsilon) = \begin{cases} \exp\left(-\varepsilon/2 \Big/ (\varepsilon_i^2/4)-r^2\right), & r < \dfrac{\varepsilon_i}{2} \\ 0 & , \; r > \dfrac{\varepsilon_i}{2} \end{cases}$$

where r is the distance from $\partial N_{\varepsilon_i} \cap \Omega_i$, could serve as the mollifier function. If we consider a sequence of numbers $\varepsilon_k = \frac{1}{2^k}$, $k = i, i+1, i+2, \ldots$ and the corresponding sequence of optimal controls $\{\hat{\phi}_k\}$, we are assured by Lemma 1 of Appendix 1 of [17] that there exists a control $\hat{\phi}$, such that some subsequence of controls $\{\phi_k\}$ (say $\{\phi_{km}\}$) converges to $\hat{\phi}$ as $k \to \infty$. Moreover,

$$\iint_\Omega \left(-\hat{\phi}\,\frac{\partial w_H}{\partial t}\right) \geq \iint_\Omega \left(-f_i\,\frac{\partial w_H}{\partial t}\right)$$

(*) is valid for any admissible control f_i acting on Ω_i. A serious problem arises at this point of the proof. Namely, we observe that while f_i is an admissible control on Ω_i, i.e.

$||f_i||_{\Omega_i} \leq 1$, it may be an inadmissible control on Ω, or even on Ω_{i+1}.

We recollect however, that either the applied loads and moments are given by a finite number of Dirac delta functions and their derivatives applied at finite number of points in Ω, in which case there exists $\tilde{\varepsilon}_1 > 0$, and a neighborhood N_{ε_1} of $\partial\Omega$ which is free from such loads, or that they are square integrable functions in Ω, bounded in the maximum norm, so that we can choose $\varepsilon_2 > 0$ and a neighborhood N_{ε_2} of $\partial\Omega$ such that given $0 < \dot{\varepsilon} < 1$, $\iiint_{N_{\varepsilon_2} \cap \Omega} |\sum_{j=1}^{n} f_j| \, dxdy < \tilde{\varepsilon}/2^i$ (for fixed i).

Hence,

$$\iint_{N_{\varepsilon_3} \cap \Omega} |f_i| < \tilde{\varepsilon}/2^i \quad \text{if } \varepsilon_3 = \min.(\varepsilon_1, \varepsilon_2).$$

(Note: Since w_H is continuously differentiable and $\frac{\partial w_H}{\partial t} \equiv 0$ on $\partial\Omega$ we can choose $\varepsilon_4 > 0$ and a neighborhood N_{ε_4} of $\partial\Omega$ such that

$$\iint_{N_{\varepsilon_4} \cap \Omega} \left|\frac{\partial w_H}{\partial t}\right| dxdy < \tilde{\varepsilon}/2^i$$

and finally denoting by $\hat{\varepsilon} = \min.(\varepsilon_3, \varepsilon_4)$ we have also:

$$\iint_{N_{\hat{\varepsilon}} \cap \Omega} |f_i \frac{\partial w_{H_i}}{\partial t}| \, dxdy < \hat{\varepsilon}^2/2^{2i}.) \qquad (**)$$

All we need to do now is to redefine the class of admissible controls to satisfy the inequality:

$$||f_i||_{\Omega_i} \leq 1 - \tilde{\varepsilon}/2^i.$$

Now if f_i is an admissible control on Ω_i, that is if $||f_i||_{\Omega_i} < 1 - \tilde{\varepsilon}/2^i$, then after reducing the radius of the corner to $1/(2^{i+1})$, we have

$$||f_i||_{\Omega_{i+1}} \leq ||f_i||_{\Omega_i} + ||\sum f_j||_{N_{\tilde{\varepsilon}}} < 1 - \frac{\tilde{\varepsilon}}{2^i} \leq 1 - \frac{\varepsilon}{2^{i+1}} \quad ,$$

and f_{i+1} is again an admissible control. We conclude that given an optimal control $\hat{\phi}(x,y,t)$ acting on Ω, and a sequence of regions Ω_i, we can select a sequence of optimal controls $\{\phi_i\}$, each ϕ_i acting in Ω_i, such that $\hat{\phi}(x,y,t) = \lim_{i \to \infty} \phi_i(x,y,t)$.

If it were not so, then we could find $\varepsilon > 0$, such that $||\phi_i - \hat{\phi}||_\Omega > \varepsilon$, for all optimal controls ϕ_i (acting on Ω_i) for sufficiently large indices i. Regarding ϕ_i as controls on Ω (i.e. $\phi_{i(\Omega)} = \phi_i$ on Ω_i and $\phi_{i(\Omega)} = 0$ on $\Omega \setminus \Omega_i$), and remembering that each ϕ_i is an optimal control on Ω_i, we obtain an easy contradiction to the statement (**). This shows that such

$\varepsilon > 0$ can not be found, and that indeed such sequence $\{\phi_i\}$ can be selected.

The corresponding sequence w_{H_i} of displacement functions satisfying the homogeneous equation must also converge to the function $w_H(\hat{\phi}(x,y,t),x,y,t)$. To prove this statement we use the Arzela-Ascoli theorem. $\{w_{H_i}\}$ form an equicontinuous family of functions. Hence some subsequence of $\{w_{H_i}\}$ must converge to a function $w_H(x,y,t)$. Because of the elastica hypothesis concerning each function w_{H_i}, $\tilde{w}_H(x,y,t)$ is also a differentiable function. Using Duhamel's principle we have for arbitrary $t \in [0,T]$:

$$\tilde{w}_H(x,y,t) - w_H(x,y,t)$$

$$= \int_0^t \iint_\Omega G((x-\xi),(y-\eta),(t-\tau)) \; [\hat{\phi}(x,y,t) - \lim_{i\to\infty} \phi_i(x,y,t)]dxdydt = 0,$$

so that $\tilde{w}_H(x,y,t) = w_H(x,y,t)$, $t \in [0,T]$.

Now given an arbitrary admissible control f acting on Ω, we can select a sequence of admissible controls $\{f_i\}$ on Ω_i such that $\lim_{i\to\infty} f_i = f$. (The argument is identical with the preceding one.) Since for each f_i the inequality (*) is valid, we have in the limit:

$$\lim_{i\to\infty} \iint_\Omega (-\phi_i \frac{\partial w_{H_i}}{\partial t})dxdy \geq \lim_{i\to\infty} \iint_\Omega (-f_i \frac{\partial w_{H_i}}{\partial t})dxdy, \text{ and}$$

finally:

$$\iint_{\Omega} (-\hat{\phi} \frac{\partial w_H}{\partial t}) \geq \iint (-f \frac{\partial w_H}{\partial t})dxdy,$$

which was to be proved.

Example 2

Consider a semicircular plate occupying the region Ω as shown on figure 4. The plate is simply supported along the diameter Γ_2 and clamped along the entire arc Γ_1. A uniform

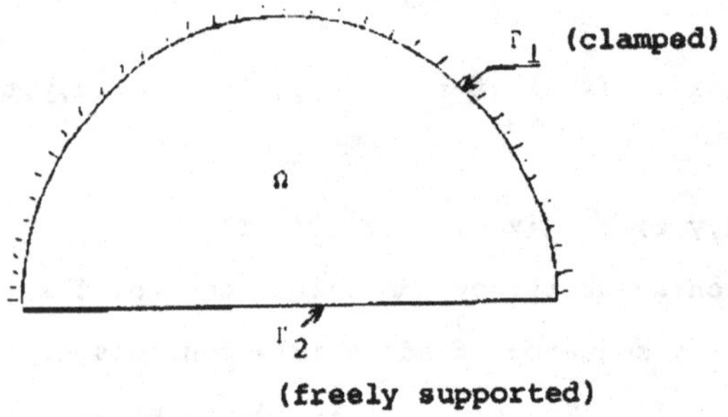

Figure 4

load p_0 is applied to this plate, then suddenly removed at the time t_0. Show that an admissible load uniformly distributed on some circular disc contained in Ω applied the instant t_0 and maintained for some time interval $[t_0,t_1]$ cannot be altered at t_1, so that the resulting control $\hat{\phi}(x,y,t)$, $t \in [t_0,T]$, $(t_1 < T)$ will be optimal on the fixed interval $[t_0,T]$, where T is chosen for convenience to be 1/4 of the time necessary to complete one cycle of vibration with the lowest natural frequency. The argument supporting this claim is similar to that of Example 1. We use the fact that the velocity of a freely vibrating plate does assume a maximum value at an isolated point in Ω. (A superposition of two uniformly distributed loads will have that effect.) Considering a Dirac delta function applied to the point of maximum velocity during a sufficiently short subinterval of $[t_0,t_1]$, we can show that the inequality (3.2.3) must be in correct in that subinterval. However, the boundary conditions stated in this example imply the validity of (3.2.3) if our assumed control load was optimal. Hence, it can not be optimal, as we intended to show.

In what follows Γ_1 will denote the part of $\partial\Omega$ obeying fixed edge condition, Γ_2 will be the simply supported part of $\partial\Omega$, and Γ_3 will denote the part of $\partial\Omega$ obeying the free edge condition.

3.2.5.3. Proof of Pontryagin's Principle for the Case When Γ_2 and Γ_3 Consists of Straight Line Intervals

For the sake of completeness we shall offer the proof of the following fundamental theorem which includes the theorems (3.2.1), (3.2.2) as special cases:

Theorem 3.2.3.

Let us assume that $\hat{\phi}(x,y,t)$ is an optimal control for the fixed time interval $[0,T]$ for a thin homogeneous plate, whose flexural rigidity and density are constant. Let the boundary consist of parts $\Gamma_1, \Gamma_2, \Gamma_3$, $(\overline{\Gamma_1 \cup \Gamma_2 \cup \Gamma_3} = \partial\Omega)$, where on Γ_1 the plate obeys the condition (B1) (i.e. clamped edge condition), on Γ_2 it obeys the condition (B2) and on Γ_3 the condition (B3). We assume that Γ_2 and Γ_3 are a union of a finite number of straight lines and that Γ_1 is a union of a finite number of piecewise smooth arcs. (Consequently that we have only a finite number of corner points.) We assume that all corners are internal corners, and all corner points are endpoints of an arc of Γ_1. Then the inequality (3.2.3) holds:

$$\iint_{\Omega} -\hat{\phi}(x,y,t) \frac{\partial w_H(x,y,t)}{\partial t} dxdy \geq \iint_{\Omega} -(f,y,t \frac{\partial w_H(x,y,t)}{\partial t} dxdy,$$

where $w_H(x,y,t)$ has the same meaning as in theorems (3.2.1) and (3.2.2).

Proof

Let us prove this theorem under the assumption that all corner points have been replaced by circular arcs of Γ_1 of sufficiently small radius, that is the assumption that $\partial\Omega$ is smooth, and then deal with the corner points in exactly the same manner as in the theorem 3.2.2. w_H denotes the displacement of the plate vibrating freely and obeying the final conditions at the time t = T:

$$w_H(x,y,T) = \hat{w}(\hat{\phi}(x,y,t),T)$$

$$v_H(x,y,T) = \frac{\partial w_H(x,y,T)}{\partial t} = \frac{\partial \hat{w}(\hat{\phi}(x,y,t,T)}{\partial t} .$$

Let $t_0 \in [0,T]$ be a point of continuity for the function $\hat{\phi}(x,y,t)$ (which may be a generalized function of x,y but is piecewise continuous, bounded function of t,) for any fixed $x,y \in \Omega$). There exists $\delta > 0$ such that $\hat{\phi}(x,y,t)$ is continuous in the interval $I_\delta = [t_0-\delta,t_0+\delta]$, and that I_δ is contained in [0,T]. Let ψ be any admissible control, such that $||\psi|| > 0$ in I_δ, and for some $\varepsilon_0 > 0$, $\hat{\phi} + \varepsilon\psi$ is an admissible control on I_δ whenever $|\varepsilon| < \varepsilon_0$. (We note that if no such ψ can be found then $\hat{\phi}$ is the only admissible control in I_δ and there is nothing to prove.)

We consider the control

$$\phi'(x,y,t) = \begin{cases} \hat{\phi}(x,y,t) \text{ for } t \in [0,T] \setminus I_\delta \\ \hat{\phi} + \varepsilon\psi \quad \text{for } t \in I_\delta . \end{cases}$$

$\phi'(x,y,t)$ is clearly an admissible control.

$\phi_\delta(x,y,t)$ will be the control:

$$\phi_\delta(x,y,t) = \begin{cases} \phi'(x,y,t) & \text{for } t \in I_\delta \\ 0 & \text{for } t \notin I_\delta, \text{ where} \end{cases}$$

δ can be chosen sufficiently small so that ϕ_δ is smooth in I_δ. Let $\hat{w}(x,y,t)$ denote the solution of MBVP corresponding to the control $\hat{\phi}(x,y,t)$ with the given boundary and initial conditions, and let $w_\delta(x,y,t)$ be the solution corresponding to the control ϕ_δ with the same boundary conditions, but with the <u>zero initial conditions</u>, (i.e. $w(x,y,0) \equiv 0$ in Ω, and $v(x,y,0) \equiv 0$ in Ω). Then the solution of MBVP corresponding to the control ϕ' is:

$$w'(x,y,t) = \hat{w}(x,y,t) + \varepsilon w_\delta(x,y,t), \tag{3.7}$$

$|\varepsilon| < \varepsilon_0$. The total energy of the plate is given by

$$\mathcal{E}(w'(x,y,t),t) = \mathcal{E}(\hat{w}(x,y,t),t) + 2\varepsilon \langle \hat{w}, w_\delta \rangle + \varepsilon^2 \mathcal{E}(w_\delta).$$

Since $\hat{w}(x,y,t)$ is an optimal displacement, we must have

$$\mathcal{E}(\hat{w},(x,y,t),T) \leq \mathcal{E}(w'(x,y,t),T)$$

independently of our choice of ε. Hence for sufficiently small ε, we must have $\langle \hat{w}, w_\delta \rangle_{(t=T)} \geq 0$. Since ι was arbitrary, and since the total energy is a continuous function of time, there must exist an interval $[T-\theta, T]$ such that

$$(\hat{w}(x,y,t) \leq (w'(x,y,t),t) \quad \text{for all } t \in [T-\theta,T].$$

Since by assumption we have $w_H(t=T) = \hat{w}(t=T)$ and $v_H(t=T) = \hat{v}(t=T)$, it must be true that

$$\langle w_H, w_\delta \rangle_{t=T} = \langle \hat{w}, w_\delta \rangle_{t=T} \geq 0.$$

Since w_δ is the displacement caused by the control ψ_δ with zero initial conditions, we must have $w_\delta \equiv 0$ in the interval $[0, t_0-\delta]$, and consequently $\langle w_H, w_\delta \rangle \equiv 0$ for all $t \in [0, t_0-\delta]$. Hence it is certainly true that $\langle w_H, w \rangle \geq 0$ in the interval $[0, t_0-\delta]$. In the interval $[t_0+\delta, T]$ the displacement function is a solution of the homogeneous equation ($A3^a$). Hence for any $\tau \in [t_0+\delta, T]$,

$$\langle w_H, w_\delta \rangle_{t=\tau} = \langle w_H, w_\delta \rangle_{t=t_0+\delta} + \int_{t_0+\delta}^{\tau} \{\frac{d}{dt} \langle w_H, w_\delta \rangle \}\, dt$$

$$= \langle w_H, w_\delta \rangle_{t=t_0+\delta} + \frac{1}{2} \int_{t_0+\delta}^{\tau} \int_{\partial\Omega} \{-v_H \frac{\partial}{\partial n}(D\nabla^2 w_\delta)$$

$$- v_\delta \frac{\partial}{\partial n} (D\nabla^2 w_H) + D\nabla^2 (w_H) \frac{\partial v_\delta}{\partial n} + D(\nabla^2 w_\delta) \frac{\partial v_H}{\partial n} \}dsdt.$$

On Γ_1 part of the boundary we have:

$$v_H = v_\delta = \frac{\partial v_H}{\partial n} = \frac{\partial v_\delta}{\partial n} \equiv 0.$$

On Γ_2 part of the boundary we have $v_H = v_\delta = D\nabla^2 w_H = D\nabla^2 w_\delta \equiv 0$ (because of our assumption that $k \equiv 0$ on Γ_2) and on Γ_3 part of the boundary we have:

$$Q_{n\delta} = \frac{\partial}{\partial n} (D\nabla^2 w_\delta) = Q_{nH} = \frac{\partial}{\partial n} (D\nabla^2 w_H) \equiv 0$$

(because of our assumption D = constant), and we also claim that:

$$D\nabla^2 w_H = D\nabla^2 w_\delta \equiv 0.$$

$Q_{nn} = 0$, and $k = 0$ on Γ_3, according to the formula ($B3^a$) implies that:

$$\frac{\partial^2 w}{\partial n^2} = - \nu \frac{\partial^2 w}{\partial s^2} ,$$

or that

$$\nabla^2 w = (1-\nu) \frac{\partial^2 w}{\partial s^2} \quad .$$

Hence

$$D[\nabla^2 w + (1-\nu) \frac{\partial^2 w}{\partial s^2}] = 2D\nabla^2 w \ ,$$

And

$$\frac{\partial}{\partial n} [D\nabla^2 w + (1-\nu) \frac{\partial^2 w}{\partial s^2}] = 0$$

(by formula B3[a], after substituting k = 0 and D = constant, and assuming that

$$\frac{\partial}{\partial s} (\frac{\partial^2 w}{\partial n \partial s}) = \frac{\partial}{\partial n} (\frac{\partial^2 w}{\partial s^2}) \ ,$$

we obtain the desired result:

$$D\nabla^2 w \equiv 0 \text{ on } \Gamma_3.$$

So finally, we have:

$$\langle w_H, w_\delta \rangle_{t=\tau} = \langle w_H, w_\delta \rangle_{t=t_0+\delta} \quad .$$

for any $\tau \in [t_0+\delta, T]$, and $\langle w_H, w_\delta \rangle_\tau = \langle w_H, w_\delta \rangle_{t=t_0+\delta} =$

$\langle w_H, w_\delta \rangle_{t=T} \geq 0.$

In the interval $I_\delta: [t_0-\delta, t_0+\delta]$ we can use the continuity of total energy as a function of time, and the Cauchy-Schwartz inequality:

$$(\langle w_H, w_\delta \rangle)^2 \leq \langle w_H, w_H \rangle \quad \langle w_\delta, w_\delta \rangle = \mathcal{E}_H \mathcal{E}_\delta$$

to conclude that

$$\lim_{\delta \to 0} \frac{1}{\delta} \int_{t_0-\delta}^{t_0+\delta} \langle w_H, w_\delta \rangle dt = \lim_{\delta \to 0} \frac{1}{\delta} \int_{t_0-\delta}^{t_0+\delta} \iint_\Omega [-\phi_\delta \frac{\partial w_H}{\partial t}] dxdydt$$

$$= 0 \qquad (*)$$

uniformly. (We recall that ϕ_δ is a smooth function of time in I_δ, and so is $\partial w_H/\partial t$.) Hence, for a sufficiently small interval of length 2δ, we must retain a constant sign of the expression $\langle w_H, w_\delta \rangle$ in I_δ. Consequently $\langle w_H, w_\delta \rangle \geq 0$ in the entire interval $[0,T]$. (We note that as before the contour integral vanished in the equation (*)). Using the relationship

$$\langle \hat{w}, w_H \rangle_{t=\tau} = \langle \hat{w}, w_H \rangle_{t=0} + \int_0^\tau \iint_\Omega (\hat{\phi} \frac{\partial w_H}{\partial t}) dxdydt, \quad (0 \leq \tau \leq T),$$

and modifying $\hat{\phi}$ on I_δ: $\tilde{\phi} = \hat{\phi} + \phi_\delta$, we obtain

$$\iint_{\Omega}(-\phi \frac{\partial w_H}{\partial t})dxdy \leq \iint_{\Omega} (-\hat{\phi} \frac{\partial w_H}{\partial t})dxdy.$$

Since I_δ was arbitrary, and since any piecewise smooth admissible control can be obtained by such modifications (within a set of measure zero) on suitable intervals I_δ, we have the required result:

$$\iint_{\Omega} (-\hat{\phi} \frac{\partial w_H}{\partial t})dxdy = \max \iint_{\Omega} (-\phi \frac{\partial w_H}{\partial t})dxdy$$

for all admissible controls ϕ and for all $t \in [0,T]$.

To extend this result to the case stated in this theorem where corner points may be present on $\partial\Omega$, we repeat the entire argument of the theorem 3.2.2. We emphasize again the fact that the corner points were restricted to be either the points which lie on Γ_1, or the points which are the end points of arc Γ_1. The limiting process which was used in the proof of the Theorem (3.2.2) succeeded because each contour integral vanished independently of the curvature k. This would not be the case if the free edge, or if the simple support conditions were present, since in that case the value of each integral would depend on k, and as the sequence of smooth boundaries approximated $\partial\Omega$, k would increase without bounds, and we would have to consider in the limit a singular contour integral, so that the conclusion reached in the theorem (3.2.2) may be incorrect.

3.2.5.4. The Case When $\partial\Omega = \Gamma_2$ and is Composed of a Finite Number of Smooth Arcs With No Corner Points

(D = const., ρ = const.)

The absence of corner points allows us to avoid singular contour integrals. However, the crucial relationship

$$\frac{d}{dt} \langle w_1, w_2 \rangle = \frac{1}{2} \iint_{\Omega} (v_2 f_1 + v_1 f_2)\,dxdy$$

is no longer true. Instead we must consider the formula (1.35[c]), modified by putting $v_1 = v_2 \equiv 0$ on $\partial\Omega$:

$$\frac{d}{dt} \langle w_1, w_2 \rangle = \frac{1}{2} \iint_{\Omega} (v_2 f_1 + v_1 f_2)\,dxdy + \frac{1}{2} \iint_{\partial\Omega} D[\nabla^2 w_1 \frac{\partial v_2}{\partial n} + \nabla^2 w_2 \frac{\partial v_1}{\partial n}]\,ds. \qquad (3.2.8)$$

If w_1, w_2 are solutions of the homogeneous equation (A3[a]) in some interval I, then the sign of $\frac{d}{dt} \langle w_1, w_2 \rangle$ is the same as the sign of

$$\int_{\partial\Omega} (D(\nabla^2 w_1 \frac{\partial v_2}{\partial n} + \nabla^2 w_2 \frac{\partial v_1}{\partial n})\,ds.$$

Since the plate is simply supported on $\partial\Omega$, we can affect some simplifications of the formula (3.2.8). We have:

$$\chi = M_{nn} + M_{\tau\tau} = -D(1+\nu)\nabla^2 w = -D(1+\nu)\left(\frac{\partial^2 w}{\partial n^2} + \frac{\partial^2 w}{\partial \tau^2}\right)$$

(see (5^a), (5^b), (5^c)). However, on $\partial\Omega$ we have:

$$w = \frac{\partial w}{\partial s} = \frac{\partial^2 w}{\partial s^2} \equiv 0, \text{ and } M_{nn} \equiv 0,$$

because of the simple support condition (B2). We use the relationship

$$\frac{\partial^2 w}{\partial \tau^2} = \frac{\partial^2 w}{\partial s^2} + k\,\frac{\partial w}{\partial n}$$

where as before k is the curvature of the boundary. Hence,

$$\chi = M_{\tau\tau} = -D(1+\nu)\left(\frac{\partial^2 w}{\partial n^2} + \frac{\partial^2 w}{\partial s^2} + k\,\frac{\partial w}{\partial n}\right)$$

$$= -D(1+\nu)\left(\frac{\partial^2 w}{\partial n^2} + k\,\frac{\partial w}{\partial n}\right) \quad \text{on } \partial\Omega \,.$$

Since $M_{nn} = 0$ on $\partial\Omega$, we also have:

$$M_{nn} = -D\left(\frac{\partial^2 w}{\partial n^2} + \nu\,\frac{\partial^2 w}{\partial \tau^2}\right) = 0,$$

or

$$\frac{\partial^2 w}{\partial n^2} = -\frac{\partial^2 w}{\partial \tau^2} = -\nu\left(\frac{\partial^2 w}{\partial s^2} + k\frac{\partial w}{\partial n}\right) = -\nu k\frac{\partial w}{\partial n}\,. \qquad (3.2.8^a)$$

Hence, on $\partial\Omega$ we obtain:

$$\chi = M_{\tau\tau} = -D(1+\nu)(1-\nu)k\frac{\partial w}{\partial n} = -D(1-\nu^2)k\frac{\partial w}{\partial n}\,. \qquad (3.2.9)$$

Using the relationship

$$\chi = -D(1+\nu)\nabla^2 w,$$

or

$$\nabla^2 w = -\frac{\chi}{D(1+\nu)}\,, \qquad (D > 0), \qquad (3.2.9^a)$$

we have on $\partial\Omega$:

$$\nabla^2 w = \frac{1-\nu^2}{1+\nu}\,k\,\frac{\partial w}{\partial n} = (1-\nu)k\,\frac{\partial w}{\partial n}\,; \qquad (3.2.10)$$

(which, of course, could be obtained directly:

$$\nabla^2 w = \frac{\partial^2 w}{\partial n^2} + \frac{\partial^2 w}{\partial \tau^2} = \frac{\partial^2 w}{\partial n^2} + k\,\frac{\partial w}{\partial n}$$

$$= -\nu k\,\frac{\partial w}{\partial n} + k\,\frac{\partial w}{\partial n} = (1-\nu)\,k\,\frac{\partial w}{\partial n}\,.)$$

Hence, the equation (3.2.8) can be rewritten:

$$\frac{d}{dt} \langle w_1, w_2 \rangle = \int_{\partial\Omega} D(1-\nu)k[\frac{\partial w_1}{\partial n}\frac{\partial v_2}{\partial n} + \frac{\partial v_1}{\partial n}\frac{\partial w_2}{\partial n}]ds$$

$$= D(1-\nu)\int_{\partial\Omega} k\frac{d}{dt}(\frac{\partial w_1}{\partial n}\frac{\partial w_2}{\partial n})ds. \qquad (3.2.8^b)$$

We are now ready to repeat the arguments of the Theorem 3.2.3. Let $w_1 = \hat{w}$ be the optimal displacement corresponding to the optimal control $\hat{\phi}(x,y,t)$. Let w_H be the solution of the homogeneous equation ($A3^a$) with the property:

$$\underset{\sim}{w}_H(x,y,T) = \hat{\underset{\sim}{w}}(x,y,T),$$

and as before we denote by $w'(x,y,t) = \hat{w} + \epsilon w_\delta$ where w_δ is a function whose support is the time interval I_δ, with the properties identical to the function w_δ described in Theorem 3.2.3. As before we have the inequality (for a suitably small ϵ_0):

$$\mathcal{E}(w') = \mathcal{E}(\hat{w}) + 2\epsilon \langle \hat{w}, w_\delta \rangle + f(\epsilon^2), \ |\epsilon| < \epsilon_0.$$

Hence, if w_H is a solution of the homogeneous equation ($A3^a$) satisfying

$$w_H(T) = \hat{w}(T)$$

$$v_H(T) = \hat{v}(T)$$

then

$$\langle w_H, w_\delta \rangle_{t=T} \geq 0.$$

But

$$\langle w_H, w_\delta \rangle_{t=T} = \langle w_H, w_\delta \rangle_{t=t_0+\delta} + \int_{t_0+\delta}^{T} \frac{d}{dt} \langle w_H, w_\delta \rangle$$

$$= \langle w_H, w_\delta \rangle_{t=t_0+\delta} + \frac{1}{2}D(1-\nu) \int_{t_0+\delta}^{T} \{ \int_{\partial\Omega} \frac{\partial}{\partial t}(k \frac{\partial w_H}{\partial n} \frac{\partial w}{\partial n}) ds \} dt$$

$$= \frac{1}{2} \int_{t_0-\delta}^{t_0+\delta} [\iint_{\Omega} (\phi_\delta v_H) dxdy] dt$$

$$+ \frac{1}{2}D(1-\nu) \int_{t_0-\delta}^{T} [\int_{\partial\Omega} \frac{\partial}{\partial t}(k \frac{\partial w_H}{\partial n} \frac{\partial w_\delta}{\partial n}) ds] dt.$$

Hence, we must have:

$$\int_{t-\delta}^{t+\delta} [\iint_{\Omega} (-\hat{\phi}_\delta v_H) dxdy] dt \leq \int_{t-\delta}^{T} [\int_{\partial\Omega} \frac{\partial}{\partial t} (k \frac{\partial w_\delta}{\partial n} \frac{\partial w_H}{\partial n}) ds] dt.$$

By an argument analogous to 3.2.3, we finally obtain for an arbitrary control $\tilde{\phi} = \hat{\phi} + \phi_\delta$ the result:

$$\int_{t_0-\delta}^{t_0+\delta} [\iint_\Omega (-\tilde{\phi} v_H) dxdy] dt \le \int_{t_0-\delta}^{t_0+\delta} [\iint_\Omega (-\hat{\phi} v_H) dxdy] dt$$

$$+ \int_{t_0-\delta}^{T} [\int_{\partial\Omega} \frac{\partial}{\partial t}(k \frac{\partial (\tilde{w}-\hat{w})}{\partial n} \frac{\partial w_H}{\partial n}) ds] dt.$$

Since δ was arbitrary, and since $\lim_{\delta \to 0} \frac{1}{\delta} \int_{t-\delta}^{t+\delta} (\bar{\phi} v_H) = 0$ uniformly, for any admissible control $\tilde{\phi}$, we have for any $t \in [0,T]$

$$\iint_\Omega (-\tilde{\phi} v_H) dxdy - \int_t^T [\int_{\partial\Omega} \frac{\partial}{\partial t} (k \frac{\partial \tilde{w}}{\partial n} \frac{\partial w_H}{\partial n}) ds] dt \le \iint_\Omega (-\hat{\phi} v_H) dxdy$$

$$- \int_t^T [\int_{\partial\Omega} \frac{\partial}{\partial t} (k \frac{\partial \hat{w}}{\partial n} \frac{\partial w_H}{\partial n}) ds] dt \qquad (3.2.10)$$

which is a form of the maximum principle of Pontryagin. It reduces to the formula (3.2.3) if we either change the boundary conditions, or if we put $k \equiv 0$ on $\partial\Omega$, or if we demand that for some reason:

$$\int_{\partial\Omega} k (\frac{\partial w}{\partial n} \frac{\partial w_H}{\partial n}) ds = \text{constant} , \qquad (3.2.11)$$

for all $t \in [0,T]$ and for any admissible displacement $w(x,y,t)$. To make the formula (3.2.10) useful we need to investigate the following problem.

What physically important criteria would assure the condition (3.2.11) for a simply supported boundary? In its present form the inequality (3.2.10) appears to be quite useless if computations of optimal control are considered. Analogous formula can be easily developed for a boundary consisting of the arcs $\Gamma_1, \Gamma_2, \Gamma_3$ obeying the boundary conditions (B1), (B2), and (B3) respectively. These formulae will not be reproduced here, since their usefulness is also questionable.

3.2.5.5.

The case when $\partial\Omega = \Gamma_1 \cup \Gamma_2$ and Γ_2 is composed of straight line segments, $\partial\Omega$ may contain internal corners which are situated anywhere on $\partial\Omega$. (As before we assume that $\partial\Omega$ is a union of a finite number of smooth arcs.) A special case when the corner points occur either on Γ_1 or at a point where an arc of Γ_1 joins an arc of Γ_2 has been already covered. We only need to consider the behavior of the line integral along some subset γ of Γ_2, which contains an interior corner.

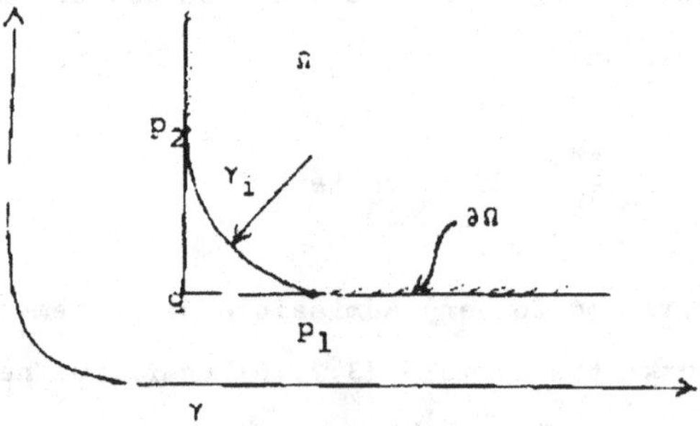

Figure 5

As in the theorem 2.2 we can approximate each corner by a sequence of circular arcs γ_i of radius $\varepsilon_i = \frac{1}{2^i}$, $i \geq N$, where N is chosen so that the circular arc ε_N lies entirely in Ω.

The contour integral (from p_1 to p_2 along γ

$$\int_\gamma k \left(\frac{\partial w}{\partial n} \frac{\partial w_H}{\partial n} \right) ds$$

can not be evaluated directly since neither k nor $\frac{\partial w}{\partial n}$ are defined at the corner point q. However, along each circular arc γ_i, we have:

$$\int_{\gamma_i} \left(k_i \frac{\partial w}{\partial n} \frac{\partial w_H}{\partial n}\right) ds = \int_{\gamma_i} \frac{1}{k_i} \left(\frac{\partial^2 w}{\partial n^2} \frac{\partial^2 w_H}{\partial n^2} \right) ds$$

because of formula (3.2.8); or using the formula (3.2.9), we have:

$$k_i \frac{\partial w}{\partial n} = -\frac{M_{\tau\tau}}{D(1-\nu^2)} ,$$

and therefore

$$\int_{\gamma_i} \left(k_i \frac{\partial w}{\partial n} \frac{\partial w_H}{\partial n}\right) ds = \frac{1}{D^2(1-\nu^2)^2} \int_{\gamma_i} \left(\frac{1}{k_i} M_{\tau\tau} M_{\tau\tau_H} \right) ds$$

$$\leq \int_{\gamma_i} \left|\frac{1}{k_i}\right| ds \cdot \int_{\gamma_i} |M_{\tau\tau}| ds \cdot \int_{\gamma_i} |M_{\tau\tau_H}| ds .$$

Since $\lim_{i\to\infty} \frac{1}{k_i} = 0$, and by assumption $\int_{\gamma} |M_{\tau\tau}| ds$ and $\int_{\gamma} |M_{\tau\tau_H}| ds$ are bounded, we obtain the desired result:

$$\lim_{i\to\infty} \int_{\gamma_i} (k_i \frac{\partial w}{\partial n} \frac{\partial w_H}{\partial n}) ds = 0.$$

The following result is an easy consequence: In the theorem (3.2.3) the last sentence, namely: "all corner points are the end points of an arc of Γ_1" can be omitted.

3.3 Instantly Optimal Controls of Thin Vibrating Plates

The definition of an instantly optimal control was given by the author in [17].

We can prove (see [17]) that if the initial fixed interval [0,T] optimal control $\hat{\phi}(x,y,t)$ satisfies the maximum principle of Theorems 3.2.1, 3.2.2, 3.2.3, i.e. if

$$\iint_{\Omega} [-\hat{\phi}(x,y,t)\frac{\partial w_H(x,y,t)}{\partial t}] dxdy \geq \iint_{\Omega} [-f(x,y,t)\frac{\partial w_H(x,y,t)}{\partial t}] dxdy$$

for any admissible control $f(x,y,t)$, then the instantly optimal control $\tilde{\phi}$ will satisfy the maximum principle

$$\iint_\Omega -[\tilde{\phi}(x,y,t)\frac{\partial w(\bar{\phi};x,y,t)}{\partial t}]dxdy \geq \iint_\Omega [-f(x,y,t)\frac{\partial w(f;x,y,t)}{\partial t}]dxdy$$

for any admissible control f.

The usefulness of this maximum principle greatly depends on the following lemma.

LEMMA 3.3.1 The instantly optimal control ϕ is unique; (that is, independent of either the manner in which we subdivided the energy, or of our choice of the intermediate optimal controls $\phi_{i,j}(x,y,t)x)$. See theorem 3 of chapter 1.

3.4 Some Comments on the Optimum Excitation Problem

We consider the following problem. (a) Let the boundary conditions of the type (B1), (B2), (B3) and the initial conditions (C1), (C2) be given for the MBVP. Find an admissible control $\hat{\phi}(x,y,t)$ for the fixed interval [0,T] such that the total energy of the plate $\mathcal{E}(T) = \mathcal{E}(\hat{\phi}(x,y,t),T)$ at the time T attains the maximum possible value, i.e., $\mathcal{E}(\bar{\phi}(x,y,t),T) \geq \mathcal{E}(f(x,y,t),T)$ for any admissible control f(x,y,t).

This problem is closely related to the resonance problem and the corresponding maximum principle reveals a physical interpretation of one possible kind of resonance. In fact any control $\phi(x,y,t)$ such that $\lim_{t\to\infty} \mathcal{E}(\phi(x,y,t),t) = \infty$ can be designated as a control of the resonance type.

A different optimal excitation is obtained by requiring a

control of MBVP to obey one of the following two conditions:

(b) The rate of increase of total energy is maximized, i.e., $d\bar{\mathcal{E}}(\bar{\phi}(x,y,t),t)/dt \geq d\mathcal{E}(f(x,y,t),t)/dt$ for any admissible control $f(x,y,t), t > 0$.

(c) Given any $\tilde{\mathcal{E}} > \mathcal{E}(t=0)$ find a control $\hat{\phi}(x,y,t)$ such that the plate attains the total energy level $\tilde{\mathcal{E}}$ in the shortest possible time.

A control satisfying (a) will be called an optimal excitation for a fixed time interval. Condition (b) will be called an excitation with the steepest rate of energy increase. Condition (c) will be called the time optimal excitation. Other definitions of optimality can be readily proposed.

To see the basic relationship between controls of the type (a) and (c) we need the following lemma.

<u>LEMMA 4.1.</u> Let the boundary conditions and the initial conditions (at t = 0) be given. Let us assume no energy transfer at the boundary $\partial\Omega$. Then given $t_1 > 0$, there exists a control $\tilde{\phi}(x,y,t)$ such that $\mathcal{E}(\phi(x,y,t),t_1) > \mathcal{E}(t=0)$.

Proof. If the initial conditions are $w(x,y,0) \equiv 0$ in Ω, (*), then any control function $\phi(x,y,t)$ such that $\phi(x,y,t) > 0$ in Ω and in a sufficiently small subinterval of $[0,t_1]$ and $\phi(x,y,t) \equiv 0$ in the remainder of $[0,t_1]$ will serve our purpose. If the initial conditions are different from (*) then there must be some subinterval $[\tau_1,\tau_2]$ of $[0,t_1]$ such that in some open neighborhood $N_{(\xi,\eta)}$ of a point $(x=\xi, y=\eta) \in \Omega$ the velocity $dw(x,y,t)/dt$ retains a constant sign. Then we apply the control

$$\phi(x,y,t) \begin{cases} \equiv 0 & , \text{ if } t \notin [\tau_1,\tau_2], \\ -\delta(x-\xi,y-\eta)\cdot \text{sign}\,\frac{dw(\xi,\eta,t)}{t} & , \text{ if } t \in [\tau_1,\tau_2], \end{cases}$$

ϕ is easily shown to increase the energy of the plate.

<u>LEMMA 4.2</u>. Every optimal excitation for a fixed time interval is also time optimal excitation.

<u>Proof.</u> We assume that there can be found $\hat{\phi}(x,y,t)$ which is an optimal excitation for the fixed time interval $[0,T]$, but fails to be a time optimal excitation, and we shall show that this assumption leads to a contradiction. Since $\hat{\phi}(x,y,t)$ was not a time optimal excitation, there must exist a control $\phi_1(x,y,t)$ such that the energy level $\hat{\mathcal{E}}(\hat{\phi}(x,y,t),T)$ can be reached in time $t_1 < T$, i.e., $\mathcal{E}(\phi_1(x,y,t),t_1) = \hat{\mathcal{E}}(\hat{\phi}(x,y,t),T)$. By the result of Lemma 4.1 there exists some admissible control $\hat{\phi}_2(x,y,t)$ on the time interval $[t_1,T]$ such that

$$\mathcal{E}(\tilde{\phi}_2(x,y,t),T) > \hat{\mathcal{E}} .$$

The control

$$\tilde{\phi} = \begin{cases} \phi_1(x,y,t), & 0 < t \leq t_1 \\ \phi_2(x,y,t), & t_1 < t \leq T, \end{cases}$$

is an admissible control, and we have

$$\mathcal{E}(\tilde{\phi}(x,y,t),T) > \hat{\mathcal{E}}(\hat{\phi}(x,y,t),T)$$

which contradicts the fact that $\hat{\phi}(x,y,t)$ was optimal for the fixed time interval $[0,T]$.

3.4.1. Pontryagin's Principle for the Optimal Excitation of a Plate for a Fixed Time Interval

Let the boundary conditions be those of either Theorem 3.2.1, or 3.2.2 or 3.2.3. Let $\bar{f}(x,y,t)$ be an optimal excitation of the plate for a fixed time interval $[0,T]$. Let $f(x,y,t)$ be any admissible control. Then the inequality

$$\iint_{\Omega} \left(\bar{f}(x,y,t) \frac{\partial w_H(x,y,t)}{\partial t} \right) dxdy \geq \iint_{\Omega} \left(f(x,y,t) \frac{\partial w_H(x,y,t)}{\partial t} \right) dxdy$$

holds (w_H has the same meaning as before).

The proof repeats the one given for the optimum control with all inequalities reversed.

<u>Some remarks concerning (3.4.1).</u> Despite the fact that this formula is identical except for the reversal of the inequality sign, with the optimal control formula, it is less useful because of the shortcomings discussed in §3.2.4. In particular, the absence of a convexity lemma is a critical defect, preventing a parallel development.

These remarks do not imply that either there is no optimal excitation for a fixed time interval [0,T], or that in some circumstances such optimal excitation could not be unique. One could artificially contruct an example of such unique excitation by starting with say circular plate with zero final deflection, and with axially symmetric final velocity distribution (of class C^J) attaining a maximum at the center, and apply maximality principle vibrating "backward it in time". For sufficiently small interval of time the optimal excitation is uniquely defined, provided our velocity distribution was suitably chosen. Now we can claim that the "backward" condition so attained is in fact the initial condition given, and we can exhibit proudly our unique optimal excitation for the just derived "fixed" time interval. Unfortunately we have no understanding of the couses of non-uniqueness, except in the instantly optimal excitation case. We shall however omit the discussion of the instantly optimal case until it becomes clear that such results are either of theoretical or of practical importance.

CHAPTER IV

Classification of the Boundary Conditions in Optimal Control Theory of Beams and Thin Plates

4.0. Introduction

It was shown that for the so-called "standard" boundary conditions of the vibrating beam (i.e. each end is either simply supported, free, or built in), Pontryagin's maximal principle takes the form:

$$- \int_{-\ell/2}^{+\ell/2} \tilde{f}(x,t) \frac{\partial w_H^*(x,t)}{\partial t} dx = \max \{- \int_{-\ell/2}^{+\ell/2} \phi(x,t) \frac{\partial w_H^*(\phi(x,t))}{\partial t} dx\}. \tag{4.1}$$

Here $\tilde{f}$ is an optimal control, $w_H^*(x,t)$ is the solution of the adjoint homogeneous equation satisfying the unique final conditions, ϕ is an arbitrary admissible control, $w(\phi(x,t))$ is the corresponding displacement function.

We could refer to the maximality conditions of this form as simple relative to the specified boundary conditions. Namely a maximality condition is simple if the boundary conditions do not explicitly enter into the corresponding inequality. In the thin plate theory the situation becomes somewhat confused. Here it has been shown in Chapter 3 that if the entire boundary is built in, or if a part of it is built in and a part is simply supported, but the simply supported part of the boundary consists only of straight lines with only a finite number of internal corners, then the maximum principle assumes a simple form. There are some easy examples of non-simple maximality conditions in thin plate theory, (see for example equations 3.2.8[b])). In this monograph we classify the boundary conditions of beam and plate theory. A separate case must be made for the coupled vibrations. It is rather surprising that a simple form of maximality principle is also correct for the coupled vibrations of beams, if the boundary conditions are of certain (physically motivated) type. However, in the general case the boundary terms will occur in the maximal principle. A physical interpretation is given for some cases of the non-simple forms of maximality principle.

4.1. The Energy Terms

In this discussion we cannot ignore the potential energy due to the effects of boundary forces. In plate theory we have the additional energy terms:

$$U_{FB} = \oint_{\partial\Omega} [Q_n w - M_{nn} \frac{\partial w}{\partial n} - M_{ns} \frac{\partial w}{\partial s}] ds, \qquad (4.2)$$

the potential energy due to the elastic support on the boundary

$$U_{EB} = 1/2 \oint_{\partial\Omega} [K_\theta (\frac{\partial w}{\partial n})^2 + K_w w^2] ds. \qquad (4.3)$$

The total energy is given by the source of kinetic energy,

potential energy due to bending, potential energy due to membrane forces and the terms U_{FB}, U_{EB}, i.e.

$$\mathcal{E} = K + U + U_{\phi} + U_{FB} + U_{EB} \ . \qquad (4.4)$$

The effect of shear distortion due to the Reissner effect (see [22]) near the boundary will be ignored. If the shear forces and moments are expressed in terms of displacements, the expression (4.2) becomes:

$$\begin{aligned} U_{FB} = \int_{\partial\Omega} & (w\{\frac{\partial}{\partial n}[-D(\frac{\partial^2 w}{\partial n^2} + \nu(\frac{\partial^2 w}{\partial s^2} + k\frac{\partial w}{\partial n}))] \\ & + \frac{\partial}{\partial s}[-D(1-\nu)(\frac{\partial^2 w}{\partial n \partial s} - k\frac{\partial w}{\partial s})]\} \\ & - \frac{\partial w}{\partial n}\{D(\frac{\partial^2 w}{\partial n^2} + \nu(\frac{\partial^2 w}{\partial s^2} + k\frac{\partial w}{\partial n})) + (1-\nu)\frac{\partial w}{\partial n} D(\frac{\partial^2 w}{\partial n \partial s} - k\frac{\partial w}{\partial s})\})dxdy. \end{aligned} \qquad (4.5)$$

We formulate an inner product on $\partial\Omega$, analogous to Szegö product occurring in the theory of analytic functions:

$$\begin{aligned} \langle w_1, w_2 \rangle_{\partial\Omega} = \frac{1}{2}\int_{\partial\Omega} & (w_1\{\frac{\partial}{\partial n}[-D(\frac{\partial^2 w_2}{\partial n^2} + \frac{\partial^2 w_2}{\partial s^2} + k\frac{\partial w_2}{\partial n})] \\ & + \frac{\partial}{\partial s}[-D(1-\nu)(\frac{\partial^2 w_2}{\partial n \partial s} - k\frac{\partial w_2}{\partial s})]\} - \frac{\partial w_1}{\partial n} D[\frac{\partial^2 w_2}{\partial n^2} \\ & + \nu(\frac{\partial^2 w_2}{\partial s^2} + k\frac{\partial w_2}{\partial n})] + (1-\nu)\frac{\partial w_1}{\partial s}\{D(\frac{\partial^2 w_2}{\partial n \partial s} - k\frac{\partial w_2}{\partial s})\} \end{aligned}$$

$$+ w_2\{\frac{\partial}{\partial n}[-D(\frac{\partial^2 w_1}{\partial n^2} + \frac{\partial^2 w_1}{\partial s^2} + k\frac{\partial w_1}{\partial n})] + \frac{\partial}{\partial s}[-D(1-\nu)\frac{\partial^2 w_1}{\partial n \partial s} - k\frac{\partial w_1}{\partial s})]\}$$

$$- \frac{\partial w_2}{\partial n}\{D[\frac{\partial^2 w_1}{\partial n^2} + \nu(\frac{\partial^2 w_1}{\partial s^2} + k\frac{\partial w_1}{\partial n})]\} + (1-\nu)\frac{\partial w_2}{\partial s} D(\frac{\partial^2 w_1}{\partial n \partial s} - k\frac{\partial w_1}{\partial s})\})ds. \tag{4.6}$$

The equivalent form for beam theory is trivial. However, the clue to the significance of terms in expression (4.6) is obtained by examining first the corresponding expression of beam theory. Integrating by parts the product $\langle w_1, w_2\rangle$ given by formula (4.2), we have

$$\frac{d}{dt}\langle w_1, w_2\rangle = \frac{1}{2}\int_{-\ell/2}^{+\ell/2} [\dot{w}_1(\rho A\frac{\partial^2 w_2}{\partial t^2}) + \dot{w}_2(\rho A\frac{\partial^2 w_1}{\partial t^2})$$

$$+ EI(\frac{\partial^2 w_2}{\partial x^2}\frac{\partial^2}{\partial x^2}(\dot{w}_1) + \frac{\partial^2 w_1}{\partial x^2}\frac{\partial^2}{\partial x^2}(\dot{w}_2))]dx, \tag{4.7}$$

(where $\cdot \equiv \frac{\partial}{\partial t}$).

$$\frac{d}{dt}\langle w_1, w_2\rangle = \frac{1}{2}\left(EI\frac{\partial^2 w_2}{\partial x^2}(\frac{\partial}{\partial x}\dot{w}_1) + EI\frac{\partial^2 w_1}{\partial x^2}(\frac{\partial}{\partial x}\dot{w}_2)\right)\Big|_{-\ell/2}^{+\ell/2}$$

$$- \frac{1}{2}\int_{-\ell/2}^{+\ell/2}\{\frac{\partial}{\partial x}(\dot{w}_1)\cdot\frac{\partial}{\partial x}(EI\frac{\partial^2 w_2}{\partial x^2}) + \frac{\partial}{\partial x}(\dot{w}_2)\frac{\partial}{\partial x}(EI\frac{\partial^2 w_1}{\partial x^2})\}dx$$

$$+ \frac{1}{2}\int_{-\ell/2}^{+\ell/2}[\dot{w}_1(\rho A\frac{\partial^2 w_2}{\partial t^2}) + \dot{w}_2(\rho A\frac{\partial^2 w_1}{\partial t^2})]dx$$

$$= \frac{1}{2}[EI(\frac{\partial^2 w_2}{\partial x^2}\frac{\partial \dot{w}_1}{\partial x} + \frac{\partial^2 w_1}{\partial x^2}\frac{\partial \dot{w}_2}{\partial x})]\Bigg|_{-\ell/2}^{+\ell/2}$$

$$- \frac{1}{2}[\dot{w}_1 \frac{\partial}{\partial x}(EI\frac{\partial^2 w_2}{\partial x^2}) + \dot{w}_2 \frac{\partial}{\partial x}(EI\frac{\partial^2 w_1}{\partial x^2})]\Bigg|_{-\ /2}^{+\ /2}$$

$$+ \frac{1}{2}\int_{-\ell/2}^{+\ell/2} [\dot{w}_1 f_2(x,t) + \hat{w}_2 f_1(x,t)]dx. \qquad (4.8)$$

In the case when $f_2(x,t) \equiv 0$, $w_2(x,t) = w_H$, $f_1(x,t) = f$, we can simplify the formula (4.8).

$$\frac{d}{dt}\langle w,w_H\rangle = \frac{1}{2}\int_{-\ell/2}^{+\ell/2} (f(x,t)\dot{w}_H)dx + \frac{1}{2}(M(w_H)\frac{\partial \dot{w}}{\partial x} + M(w)\frac{\partial \dot{w}_H}{\partial x})\Bigg|_{-\ell/2}^{+\ell/2}$$

$$- \frac{1}{2}(Q(w_H)\dot{w} + Q(w)\dot{w}_H)\Bigg|_{-\ell/2}^{+\ell/2} \qquad (4.9)$$

(where $M(w)$ stands for the bending moment $M(w(x)) = EI\frac{\partial^2 w}{\partial x^2}$, and Q for the shear force $Q(w(x)) = \frac{\partial}{\partial x}(EI\frac{\partial^2 w}{\partial x^2})$.) We observe that the boundary terms vanish if we assume one of the following boundary conditions: (B1), (B2), or (B3). This assumption was made in [17] in derivation of the simple form of Pontryagin's principle for the linear beam theory. In this chapter we shall retain the boundary terms and observe the consequence of alternate assumptions which do not exclude the possibility of energy transfer at the boundary.

We are now ready to repeat the argument of [12], [13], or [23], in deriving the maximality principle. Assuming piecewise continuity of $f(x,t)$, $\frac{\partial w_H}{\partial t}$ as functions of t (for a fixed x), we assume that $\phi(x,t)$ is the optimal control, i.e. ϕ is an admissib;e control such that some energy form is minimized at the time $t=T$. In the case, when we wish to minimize $\mathcal{E}(t) = K + U$

$$= \frac{1}{2}\int_{-\ell/2}^{+\ell/2} \rho A(x)\left(\frac{\partial w}{\partial t}\right)^2 dx + \frac{1}{2}\int_{-\ell/2}^{+\ell/2} EI\left(\frac{\partial^2 w}{\partial x^2}\right)^2 dx \text{ evaluated at } t = T,$$

the argument is completely analogous to [12], only occasionally slightly more complicated.

Theorem 4.1. (Existence of optimal control. There exists at least one optimal control. No changes in the proof given in chapter 2 are necessary in this case.

Theorem 4.2. (Uniqueness of the finite state.) Let $\phi_1(x,t)$, $\phi_2(x,t)$ be two optimal controls minimizing $K(t) + U(t)$, on the (fixed) interval $[0,T]$. Then $w_1(\phi_1,x,T) = w_2(\phi_2,x,T)$ and $\dot{w}_1(\phi_1,x,T) = \dot{w}_2(\phi_2,x,T)$.

Theorem 4.3. (Pontryagin's principle.) Let $\phi(x,t)$ be an optimal control for the fixed interval $[0,T]$, optimizing $\mathcal{E}(t) = U(t) + K(t)$ at the time $t = T$. Then for an arbitrary admissible control $f(x,t)$ it is true that

$$\int_{-\ell/2}^{+\ell/2} [-\phi(x,t)\dot{w}_H(x,t)]dx - \{[M(w_H)\frac{\partial \dot{w}(\phi)}{x} + M(w(\phi))\frac{\partial \dot{w}_H}{\partial x}]$$

$$- [Q(w_H)\dot{w}(\phi) + Q(w(\phi))\dot{w}(H)\}\Big|_{-\ell/2}^{+\ell/2} \geq \int_{-\ell/2}^{+\ell/2} [-f(x,t)\dot{w}_H]dx$$

$$- \{[M(w_H)\frac{\partial \dot{w}(f)}{\partial x} + M(w(f))] - [Q(w_H)\dot{w}(f) + Q(w(f))\dot{w}_H]\}\Big|_{-\ell/2}^{+\ell/2}$$

(4.10)

for almost all $t \in [0,T]$. $w_H(x,t)$ is the solution of the homogeneous equation (2.1^a) satisfying unique <u>final</u> condition (i.e. at $t = T$) established in theorem 4.2. The main arguments of the proof heve been given in Chapter 2. Modifying the optimal control as before, we obtain the inequality

$$<w_H,w(\phi)>_{(t)} + \frac{1}{2}\int_t^T \{[M(w_H)\frac{\partial \dot{w}(\phi)}{x} + M(w(\phi))\frac{\partial \dot{w}_H}{\partial x}] - [Q(w_H)\dot{w}(\phi)$$

$$+ Q(w(\phi))\dot{w}_H]\}\Big|_{-\ell/2}^{+\ell/2} dt \leq <w_H,w(f)>_t + \frac{1}{2}\int_t^T \{[M(w_H)\frac{\partial \dot{w}(f)}{\partial x}$$

$$+ M(w(f))\frac{\partial \dot{w}_H}{\partial x}] - [Q(w_H)\dot{w}(f) + Q(w(f))\dot{w}_H]\}\Big|_{-\ell/2}^{+\ell/2} dt;$$

(4.11)

for any admissible control $f(x,t)$, and corresponding displacement $w(f)$. The formula (4.11) is the maximality principle. It can be reduced to a more familiar form.

$$<w_H,w(\phi)>_{(t_1)} = \frac{1}{2}\int_{t_0}^{t_1} [\int_{-\ell/2}^{+\ell/2} (\phi(x,t)w_H)dx + \frac{1}{2}\int_{t_0}^{t_1} \{[M(w_H)\frac{\partial \dot{w}(\phi)}{\partial x}$$

$$+ M(w(\phi))\frac{\partial \dot{w}_H}{\partial x}] - [Q(w_H)\dot{w}(\phi) + Q(w(\phi))\dot{w}_H]\}\Big|_{+\ell/2}^{-\ell/2} dt$$

$$\leq \frac{1}{2}\int_{t_0}^{t_1} [\int_{-\ell/2}^{+\ell/2} (f(x,t)\dot{w}_H)dx + \frac{1}{2}\int_{t_0}^{t_1} \{[M(w_H)\frac{\partial \dot{w}(f)}{\partial x}$$

$$+ M(w(f))\frac{\partial \dot{w}_H}{\partial x}] - [Q(w_H)\dot{w}(f) + Q(w(f))\dot{w}_H]\}\Big|_{-\ell/2}^{+\ell/2} ,$$

for an arbitrary interval $[t_0,t_1]$ $[0,T]$ and for an arbitrary admissible control $f(x,t)$, such that $f(x,t) = \phi(x,t)$ on $[0,t_0]$. It follows easily that if t_0,t_1 are points of continuity and $[t_0,t_1]$, is any subinterval of $[0,T]$, on which f is continuous and the appropriate derivatives exist, then we must have:

$$\int_{-\ell/2}^{+\ell/2} (-\phi(x,t)\dot{w}_H)dx - \{[M(w_H)\frac{\partial \dot{w}(\phi)}{\partial x} + M(w(\phi))\frac{\partial \dot{w}_H}{\partial x}] - [Q(w_H)\dot{w}(\phi)$$

$$+ Q(w(\phi))w_H]\}\Big|_{-\ell/2}^{+\ell/2}$$

$$\geq \int_{-\ell/2}^{+\ell/2} (-f(x,t)\dot{w}_H)\, dx - \{[M(w_H)\frac{\partial \dot{w}(f)}{\partial x}$$

$$+ M(w(f))\frac{\partial \dot{w}_H}{\partial x}] - [Q(w_H)\dot{w}(f) + Q(w(f))\dot{w}_H]\}\Big|_{-\ell/2}^{+\ell/2}$$

(4.12)

for any admissible control $f(x,t)$, for almost all $t \in [0,T]$. This completes the proof.

It is clear that for arbitrary end conditions the maximality principle is not of the simple type (according to our definition in the introductory remarks). It is simple if for almost all $t \in [0,T]$ we have

$$\{[M(w_H)\frac{\partial \dot{w}(f)}{\partial x} + M(w(f))\frac{\partial \dot{w}_H}{\partial x}] - [Q(w_H)\dot{w}(f) + Q(w(f))\dot{w}_H]\}_{x=+\ell/2}$$

$$= \{M(w_H)\frac{\partial \dot{w}(f)}{\partial x} + M(w(f))\frac{\partial \dot{w}_H}{\partial x}] - [Q(w_H)\dot{w}(f)$$

$$+ Q(w(f))\dot{w}_H]\}_{x=-\ell/2} .$$

This is easily satisfied if the conditions ar both end points ($x = \pm\ell/2$) satisfy either(3B1) or(3B2) or(3B3). While it is possible to visualize special cases when neither of these boundary conditions are satisfied, and yet the maximality principle turns out to be of the simple type, they are clearly special cases. Say when the optimal final state turns out to be symmetric (in both deflection and velocity). In general we see that the maximality principle will be of the simple type if the boundary condition at $x = \pm\ell/2$ are (2.4a) or (2.4b) or (2.4c).

Another look at the boundary terms occuring in the inequality (4.12), and we easily interpret the physical meaning of each term. In fact the boundary terms represent the rate of energy transfer at the boundary to do the work performed by the moments and shear loads.

For convenience we shall denote by $B(w_1, w_2)$ the following energy product:

$$B(w_1,w_2) = \{M(w_1)\frac{\partial w_2}{\partial x} + M(w_2)\frac{\partial w_1}{\partial x} - Q(w_1)w_2 - Q(w_2)w_1\}_{x=+\ell/2}$$

$$- \{M(w_1)\frac{\partial w_2}{\partial x} + M(w_2)\frac{\partial w_1}{\partial x} - Q(w_1)w_2 - Q(w_2)w_1\}_{x=-\ell/2} \quad (4.13)$$

The optimal control problem can now be redefined, and we attempt to incorporate the work performed on the boundary into the statement of the control problem.

We consider the following control problem: Given the initial conditions of the type (2.5) and quite arbitrary conditions prescribed on the boundary, however, such that we are assured of the unique solution of the equation (1), find an admissible control $\phi(x,t)$ on the time interval $[0,T]$ such that

$$(w(\phi))_{t=T} - \frac{1}{2} B(w(\phi),w(\phi))_{t=T}$$

is minimal. That is for an arbitrary admissible control $\phi'(x,t)$, we have

$$(w(\phi'))_T - \frac{1}{2} B(w(\phi'),w(\phi'))_T$$

$$\geq (w(\phi))_T - \frac{1}{2} B(w(\phi),w(\phi))_T. \quad (4.14)$$

In analogy with our previous arguments we set $\phi' = \phi + \varepsilon\psi_\sigma$. The inequality (4.14) is equivalent to $2\varepsilon < w(\phi), w_\sigma>_{t=T} + \varepsilon^2 \quad (w_\sigma)_{t=T}$ $- \varepsilon\, B(w(\phi),w_\sigma)_{t=T} - \frac{1}{2}\varepsilon^2 B(w_\sigma,w_\sigma)_{t=T} > 0$. Since ε is arbitrary, this is possible only if $< w(\phi),w_\sigma>_{t=T} - \frac{1}{2} B(w(\phi),w_\sigma)_{t=T} > 0$. Since w_H attains the same final state as $w(\phi)$, the above inequality is replaced by

$$< w_H,w_\sigma>_T - \frac{1}{2} B(w_H,w_\sigma)_T > 0.$$

Since

$$\int_{\tau+\sigma}^{T} (\frac{d}{dt}< w_H,w_\sigma>)dt = \frac{1}{2}\int_{\tau+\sigma}^{T} \{M(w_\sigma)\frac{\partial\dot{w}_H}{\partial x} + M(w_H)\frac{\partial w_\sigma}{\partial x} - Q(w_\sigma)\dot{w}_H - Q(w_H)\dot{w}_\sigma\}\Big|_{-\ell/2}^{+\ell/2} dt , \qquad (4.15)$$

and

$$<w_H,w_\sigma>_T - \frac{1}{2}B(w_H,w_\sigma)_T = <w_H,w_\sigma>_{\tau+\sigma}$$
$$+ \frac{1}{2}\int_{\tau+\alpha}^{T} \{M(w_\sigma)\frac{\partial\dot{w}_H}{\partial x} + M(w_H)\frac{\partial\dot{w}_\sigma}{\partial x} - Q(w_\alpha)\dot{w}_H$$
$$- Q(w_H)\dot{w}_\sigma\}\Big|_{-\ell/2}^{+\ell/2} dt - \frac{1}{2} B(w_H,w_\sigma)_{\tau+\sigma}$$
$$- \frac{1}{2}\int_{\tau+\sigma}^{T} (\frac{d}{dt} B(w_H,w_\sigma))dt = <w_H,w_\sigma>_{\tau+\sigma}$$

$$- \frac{1}{2} B(w_H,w_\sigma)_{\tau+\sigma}$$

$$- \frac{1}{2} \int_{\tau+\sigma}^{T} \{\dot{M}(w_\sigma)\frac{\partial w_H}{\partial x} + \dot{M}(w_H)\frac{\partial w_\sigma}{\partial x} - \dot{Q}(w_\sigma)w_H$$

$$- \dot{Q}(w_H)w_\sigma\}\Big|_{-\ell/2}^{+\ell/2} dt > 0.$$

Following the same line of argument as before we obtain the maximality principle:

$$\langle w_H,w(f)\rangle_t - \frac{1}{2} B(w_H,w(f))$$

$$- \frac{1}{2} \int_{t}^{T} \{\dot{M}(w_H)\frac{\partial w(f)}{\partial x} + \dot{M}(w(f))\frac{\partial w_H}{\partial x} - \dot{Q}(w(f))w_H$$

$$- \dot{Q}(w_H)w(f)\}\Big|_{-\ell/2}^{+\ell/2} dt \geq \langle w_H,w(\phi)\rangle_t - \frac{1}{2} B(w_H,w(\phi))$$

$$- \frac{1}{2} \int_{t}^{T} \{\dot{M}(w_H)\frac{\partial w(\phi)}{\partial x} + \dot{M}(w(\phi))\frac{\partial w_H}{\partial x} - \dot{Q}(w_H)w(\phi)$$

$$- \dot{Q}(w(\phi))w_H\}\Big|_{-\ell/2}^{+\ell/2} dt \quad , \tag{4.16}$$

for all $t \in [0,T]$, for arbitrary admissible control $f(x,t)$.

Since the initial conditions at $t = 0$ are fixed, this is possible only if

$$\frac{1}{2} \int_{-\ell/2}^{+\ell/2} (f\,\dot{w}_H)dx + \frac{1}{2} \{M(w(f)\frac{\partial \dot{w}_H}{\partial x} + M(w_H)\frac{\partial \dot{w}(f)}{\partial x}$$

$$- Q(w(f))\dot{w}_H - Q(w_H)w(f)\}_{-\ell/2}^{+\ell/2}$$

$$- \frac{1}{2}\dot{B}(w(f),w_H) + \frac{1}{2}\{\dot{M}(w(f)\frac{\partial w_H}{\partial x} + \dot{M}(w_H)\frac{\partial w(f)}{\partial x}$$

$$- \dot{Q}(w(f))w_H - \dot{Q}(w_H)w(f)\}_{-\ell/2}^{+\ell/2}$$

$$= \frac{1}{2}\int_{-\ell/2}^{+\ell/2} (f\ \dot{w}_H)dx \geq \frac{1}{2}\int_{-\ell/2}^{+\ell/2} (\phi\ \dot{w}_H)dx.$$

The equivalent statement is

$$\int_{-\ell/2}^{+\ell/2} - (\phi\ \dot{w}_H)dx = \max_{f \in U} \int_{-\ell/2}^{+\ell/2} -(f\ \dot{w}_H)dx \qquad (4.17)$$

for all $t \in [0,T]$, where U denotes the set of all admissible controls.

The proof that such optimal control exists may be copied from [17] without any substantial changes in the arguments. Hence, we have a maximality principle of the simple type according to our definition. It is appropriate to make a comment concerning the reason for the simple form of the maximality principle. The energy form minimized is the true total energy of the beam, while minimization of $\mathcal{E}(t) = K(t) + U(t)$ in reality ignores the energy stored at the boundary. This supports a rather vague notion that very complicated forms of maximality principle arise out of "poorly formulated" control problems, and vice versa problems which involve extremization of basic forms of energy have usually a simple formulation of the maximality principle.

4.2. Classification of boundary conditions for control theory of thin plates

The simple form of maximality principle

The author has shown in [18] that if $D(x,y) = \text{constant}$, and if on arcs $\Gamma_1, \Gamma_2, \Gamma_3$ of the boundary $\partial\Omega$ of the plate,

$$(\overline{\Gamma_1 \quad \Gamma_2 \quad \Gamma_3} = \partial\Omega) ,$$

the boundary conditions are of the form (3B1), (3B2), (3B3) respectively (see chapter 3), and if Γ_1, Γ_3 consist of straight lines and all corners are internal corners situated either on Γ_2 or on points of intersection of $\overline{\Gamma_2}$ with either $\overline{\Gamma_1}$, or $\overline{\Gamma_3}$, then the maximality principle for the time optimal problem (see [14] is of the simple type:

$$\iint_\Omega - \phi(x,y,t)\dot{w}_H(x,y,t)\,dxdy \geq - \iint_\Omega f(x,y,t)\dot{w}_H(x,y,t)\,dxdy ,$$

where ϕ is an optimal control, and f is an arbitrary admissible control. Γ_1 is a subarc (or subarcs) of $\partial\Omega$ satisfying the fixed edge condition which will be denoted by (3B1), Γ_2 satisfies the simple support condition (3B2), Γ_3 satisfies the free edge condition (3B3). See [37] and [22] for appropriate equations. The Reissner effects and the membrane forces are ignored (see [22]).

If we do not assume $D(x,y) = $ constant, the classification problem becomes very difficult, since not only the boundary conditions, but also the geometric shape of the boundary enter into the formulation of the maximality principle. A particular case of non-simple form of maximality principle for thin plates was given in [18]. The author has refrained from detailed discussion of the formula simply because it appeared to have no valid practical applications. Again it becomes clear that the fault lies in the statement of the control problem, that is we wish to minimize physically the "wrong" form of energy. We shall now consider the following control problem: Subject to boundary conditions (3B1) on Γ_1, (3B2) on Γ_2, (3B3) on Γ_3 and to initial conditions we wish to find a control $\phi(x,y,t)$, $x,y \in \Omega$, $t \in [0,T]$ such that $\{\mathcal{E} - 1/2(B(w(\phi),w(\phi)))\}_{t=T}$ is minimized, assuming that $\min(-1/2(B(w,w))) > 0$ at the time $t = T$.

$\mathcal{E}(t) = \mathcal{E}(w(\phi(t)))$ stands for

$$\mathcal{E}(t) = \frac{1}{2}\iint_\Omega D(\nabla^2 w)^2 - D(1-\nu)\Diamond^4(w,w)\,dxdy + \frac{1}{2}\iint \rho(\dot{w})^2 dxdy,$$

and $$B(w_1,w_2) = \int_{\partial\Omega} [M_{nn}(w_1)\frac{\partial w_2}{\partial n} + M_{nn}(w_2)\frac{\partial w_1}{\partial n} + M_{ns}(w_1)\frac{\partial w_2}{\partial s}$$

$$+M_{ns}(w_2)\frac{\partial w_1}{\partial s} - Q(w_1)w_2 - Q(w_2)w_1]ds,$$

where M_{ij} are moments, Q_i are shears.

$\mathcal{E} - \frac{1}{2}B(w,w)$ is the sum of kinetic and true potential energy, i.e. the potential due to strain energy of the deflected plate and the potential energy due to the presence of boundary forces.

Following an argument identical with the corresponding beam argument of preceding section, and using the result of [18] we derive the simple form of maximality principle.

In the case of constant crossection the necessary condition for a non-zero minimum of $\mathcal{E} - 1/2(B(w,w))$ to be attained at the time $t = T$ is that the optimal control $\phi(x,y,t)$ satisfies the inequality

$$\iint_{\Omega} [-\phi(x,y,t)\dot{w}_H(x,y,t)]dxdy$$

$$\geq \iint_{\Omega} [-f(x,y,t)\dot{w}_H(x,y,t)]dxdy$$

for any admissible control $f(x,y,t)$ and for all $t \in [0,T]$.

The meaining of w_H is the same as before. In fact the simplification of formulas offered in [18] which was used in deriving this result remains correct if $D(x,y)$ is a linear function of x,y rather than constant. In general case of variable modulus of flexual rigidity D(x,y) we conjecture that the above result still remains correct, although at the present time we have not been able to complete this rather complicated manipulation of formulas.

4.3. Comments on Optimal excitation theory

A dual problem to that of optimal control is the following optimality problem which we shall call the optimal excitation problem.

We consider a similar problem with given properly posed boundary and initial conditions, and wish to find an admissible control $\phi(x,y,t)$, $t \in [0,T]$ such that the total energy $\mathscr{E}(t)$ (or some other functional) is maximized at the time $t = T$.

It is easy to retrace all steps of the optimal control theory with the appropriate energy inequalities reversed to obtain analogous results for all known theorems of optimal control theory. In particular our classification of boundary conditions is equally pertinent to this problem. However, the optimality principles for optimal excitation theory have an entirely different interpretation and in many cases it is doubtful whether they have practical value. The crucial question of what is the meaining of w_H in optimal excitation theory is hard to answer. In optimal control theory $w_H(x,y,t)$ is the solution of the homogeneous equation (of beam, or plate theory) satisfying the unique final condition at $t = T$, and the appropriate boundary conditions. In optimal excitation theory we have no uniqueness theorem available to give an analogous definition of w_H. Instead, w_H is a solution of the homogeneous equation satisfying some optimal final condition, attained by applying an optimal excitation to the beam or plate subject to given boundary conditions and to initial conditions.

Optimal controls are generally not unique. However, optimal controls form a closed convex subset of the set of admissible controls. This statement is false in the case of optimal excitations. In fact if $\phi_1(x,y,t)$, $\phi_2(x,y,t)$ are linearly independent optimal excitations then it is easily shown that $\phi = \lambda\phi_1 + (1-\lambda)\phi_2$, $0 < \lambda < 1$, can not be optimal.

These questions concerning the value of optimality principles do not affect the fact that for purposes of classification of boundary conditions our entire argument is valid if "optimal control" is replaced by "optimal excitation".

An early study of the boundary control of oscillations can be found in [4]. A more abstract treatment of boundary conditions has been given by Fattorini (see example [14]). The treatment of boundary control of symmetric hyperbolic system has been given by Russell in [33] and recently in [46]. (Also see monograph by Lions [45].)

APPENDIX A

5.1 Expository comments concerning Pontryagin's maximality principle. The "general" problem of optimal control of a system of ordinary differential equations consists of finding an admissible control (function, or generalized function) $\underset{\sim}{u}(t)$ such that the following system of differential equations is satisfied, $\dot{\underset{\sim}{x}} = f(t, \underset{\sim}{x}(t), \underset{\sim}{u}(t)), \underset{\sim}{x} \in E^n, \underset{\sim}{u} \in E^m, m \leq n$, and that some functional $I(\underset{\sim}{x}(t), \underset{\sim}{u}(t))$ assumes an extreme value. Commonly this functional I is of the form $I(\underset{\sim}{x}(t), \underset{\sim}{u}(t)) = \int_{t_0}^{T} f_0(\underset{\sim}{x}(t), \underset{\sim}{u}(t)dt$, Assigning a new coordinate

$x_0(t) = \int_{t_0}^{t} f_0(\underset{\sim}{x}(\xi), \underset{\sim}{u}(\xi))\, d\xi$, we obtain a new system of differential equations $\dot{\underset{\sim}{x}} = \underset{\sim}{f}(t, \underset{\sim}{x}(t), \underset{\sim}{u}(t))$, $\underset{\sim}{x} = (x_0, x_1, \ldots, x_n)$. Obviously $x_0(t_0) = 0$, $x_0(T) = I$. In the case $f_0 \equiv 1$ the problem is referred to as "time optimal" control problem. Let us assume that $f_0(\underset{\sim}{x}(t), \underset{\sim}{u}(t)) > 0$ for all admissible vectors $\underset{\sim}{u}(t)$, so that $x_0(t)$ is a monotone increasing function of t. The problem has now been reduced to a time optimal type control problem.

A convenient change of variable results in parametrizing the dependent variables, using x_0 instead of t as the parameter. We have

$$\frac{dt}{dx_0} = \frac{1}{f_0}(\underset{\sim}{x},\ \underset{\sim}{u}(t)) = \frac{1}{\hat{f}_0(\underset{\sim}{x}(x_0),\underset{\sim}{u}(x_0))}$$

$$\underset{\sim}{v}(x_0) = \underset{\sim}{u}(t(x_0)) \qquad (5.2)$$

$$\underset{\sim}{y}(x_0) = \underset{\sim}{x}(t(x_0)) \quad .$$

It is easy to see that $\underset{\sim}{v}(x_0)$ is again an admissible control. The new system of differential equations is

$$\frac{dy_i}{dx_0} = \frac{f_i(y,\ v)}{f_0(x(x_0),\ u(x_0))} \qquad i = 1,2,\ldots,n \quad , \qquad (5.3)$$

subject to initial condition $\underset{\sim}{y}(x_0 = 0) = \underset{\sim}{x}(0)$. The functional to be minimized is $x_0(T) = I(T)$. Pontryagin's theory introduces the dual variable $\psi_i(t)$, and the "Hamiltonian" $H(\psi_i,\ x_i,\ \underset{\sim}{u})$, satisfying "canonical" equations

$$\frac{\partial H}{\partial x_i} = -\dot{\psi}_i \quad , \qquad \frac{\partial H}{\partial \psi_i} = \dot{x}_i = f_i(\underset{\sim}{x}(t),\ \underset{\sim}{u}(t)) \quad , \qquad (5.4)$$

$$H = \sum_{i=1}^{n} \psi_i f_i \quad , \qquad H(\psi_i(T),\ f_i(T)) \geq 0 \quad .$$

Denoting by $M = \max_{u \in U} H(\psi(t),\ f(t),\ u)$ and $\psi_0 = -M$, (hence $\psi_0 \leq 0$),

Pontryagin introduces for the time optimal problem the modified "Hamiltonian" $\mathcal{H} = \sum_{i=0}^{n} \psi_i(t) f_i(\underset{\sim}{x}, \underset{\sim}{u})$. The present scheme is easily converted to the time optimal control problem along the lines suggested above, and we obtain

$$H(\zeta_i(x_0), y(x_0), v) = \sum_{i=1}^{n} \zeta_i(x_0) \frac{f_i(y, v)}{f_0(y, v)} \quad \text{where}$$

$\zeta_i(x_0) = \psi_i(t(x_0))$.

$$\mathcal{H} = \sum_{i=1}^{n} (\zeta_i(x_0) \frac{f_i(\underset{\sim}{y}(x_0), \underset{\sim}{v}(x_0))}{f_0(\underset{\sim}{y}(x_0), \underset{\sim}{v}(x_0))} + \psi_0 f_0(y, \underset{\sim}{v}) \quad . \qquad (5.5)$$

The canonical equations are

$$- \dot{\psi}_i = \frac{\partial \mathcal{H}}{\partial x_i} \quad , \quad \dot{x}_i = \frac{\partial \mathcal{H}}{\partial \psi_i} \quad , \quad i = 0,1,2,\ldots,n \quad , \qquad (5.5^{a})$$

(or their equivalent in the parameter x_0). The Pontryagin's maximality principle then asserts that the necessary condition for an admissible control u(t) to be optimal along a corresponding trajectory $\hat{x} = x(t, 0, \underset{\sim}{x}_0)$ is the existence $\psi_0 \leq 0$, and of a solution $\hat{\psi}(t)$ to the dual system of equations

$$- \dot{\hat{\psi}}_i = \sum_{i=0}^{n} \hat{\psi}_i(t) \frac{\partial f_i(\underset{\sim}{x}, \underset{\sim}{u})}{\partial x_i} \qquad (5.6)$$

such that

$$\max_{u \in U} \mathcal{H}(\psi, x, u) = \mathcal{H}(\hat{\psi}, \hat{x}, u) \equiv 0 , \qquad u \in U$$

along the entire trajectory of the system $0 \leq t \leq T$, at all points of continuity of $u(t)$. The non-autonomous systems are handled by enlarging the dimension of the space, introducing $x_{n+1} \equiv t$, i.e. $\dot{x}_{n+1} \equiv 1 = f_{n+1}$, $x_{n+1}(0) = 0$, and modifying the Hamiltonian H by adding $\psi_{n+1} = (\psi_{n+1} \cdot 1)$. Problems with constraints of the form $\int_0^T g_i(x, u)dt = b_i$, $i = 1,2,\ldots,k$, are handled by the usual Lagrange multiplier technique. For popular exposition of maximality principle of Pontryagin see for example [20]. For the original proofs see [28] and [29].

APPENDIX B

Some related problems in control theory for vibrating beams. A different type of control problem for the vibrating beam was considered by Barnes in [49]. $w(x,t,u(,t))$ will denote the displacement of the beam subject to control $u(x,t)$, $t > 0$, $x \in [0,\ell]$. The problem consists of finding an admissible control $u^0(x,t)$ which minimizes at some time $t = T$ the functional

$$J(u(x,t),T) = \int_0^\ell [g_1(x,w(x,T)) + g_2(x,\frac{\partial w}{\partial t}(x,T))]dx$$

$$+ \int_0^T\int_0^\ell [F_0(x,t),w(x,t)u(x,t)]dxdt, \text{ subject to constraints}$$

$$\int_0^\ell h_i(x,w(x,T))dx + \int_0^T\int_0^\ell F_{-i}(x,t,w(x,t)u(x,t))dxdt = C_{-i},\ i = 1,2.$$

And $\int_0^T\int_0^\ell F_i(x,tw(x,t),u(x,t))dxdt \le C_i$, $i = 1,2,\ldots,n$. There Pontryagin's maximum principle for the optimal control $u^0(x,t)$ is:

$$\max_{n\varepsilon U}[v(x,t)u(x,t) + \sum_{i=-2}^{n} \lambda_i F_i(x,tw^0(x,t)u(x,t)] = v(x,t)u^0(x,t)$$

$$+ \sum_{i=-2}^{n} \lambda_i F_i(x,t)w^0(x,t)u^0(x,t),$$ where $v(x,t)$ satisfies the modified differential equation $\rho(x)A(x)\frac{\partial^2 v}{\partial t^2} + \frac{\partial^2}{\partial x^2}(EI(x)\frac{\partial^2 v}{\partial x^2}) =$

$\sum_{i=-2}^{n} \lambda_i \frac{\partial F_i}{\partial u}(x,t,u^0,w^0(x,t))$, satisfies the given boundary conditions, and the finite conditions given at $t = T$: $v(x,T) =$

$$\frac{\lambda_0}{\rho A}\frac{\partial g_2}{\partial u}(x, \frac{\partial w^0}{\partial t}(x,T)) + \frac{\ell-2}{\rho A}\frac{\partial h_2(x,\frac{\partial w_0(x,T)}{\partial t})}{\partial u};\quad \frac{\partial v(x,T)}{\partial t} =$$

$$\frac{\lambda_0}{\rho A}\frac{\partial g_1}{\partial u}(x,w^0(x,T)) - \frac{\lambda_{-1}}{\rho A}\frac{\partial h_1(x,w^0(x,T))}{\partial u}.$$

λ_i, $i = -1,-2,0,1,\ldots,n$, are corresponding Lagranian multipliers obeying the inequalities $\lambda_i \leq 0$, $i = 1,2,\ldots,n$.

A physical interpretation of such problem arises when for example we wish to exert a >0 the shape of the beam should approximate in some (to be defined) sense in the best possible manner a given admissible shape, while the kinetic energy is to be kept below some a priori assigned value.

Finally we should mention here a numerical technique for improving a suboptimal control of a vibrating beam which is given by M. Kuehne in [50]. Kuehne uses a decoupling technique for matrix Riccati equation for the corresponding pointwise control problem. Details of this technique, and further theoretical results are to be published in the near future in an article by Gilles and Kuehne.

APPENDIX C

5.2. The class of admissible loads of beam and plate theory.

A complete discussion of the assertion made in this appendix can be found in the reference [47]. Morever the linear beam and plate theory have been singled out only for reasons of simplicity, and the arguments of [47] make no use of the specific hypothesis of thin plate, or beam theory. The results are in fact easily generalized to thin shell theory, and to other important engineering applications.

We consider the class of weak solutions of the basic plate (or beam) equation (see equations (2.1) and (2.1^a)), which are postulated to satisfy the "elastica" hypothesis, and therefore are assumed to be elements of Sobolev space W_2^2, i.e. we require for the deflection function the existence of generalized derivatives of order two (in the sense of Sobolev) where the usual and weak derivatives are elements of the space $L_2(\Omega)$. For more consistent approach perhaps both displacements ard loads could be considered as distributions in the sense of Schwartz. Then a direct application of Sobolev's lemma results in the following classification of admissible loads.

A solution of the basic plate (beam) equation satisfies the "elastica" hypothesis if and only if it is an element of a space of distributions $(K^1(\Omega))^*$ consisting of

1) Bounded, measurable functions $q(x)$ on Ω. (In this argument we do not need the additional condition

$\int |q(x)|dx \leq 1$). We shall refer to these loads as the admissible distributed loads.

2) The dirac delta function concentrated at a finite number of points of Ω.

3) The first derivative of the Dirac delta function concentrated at a finite number of points of Ω.

Hence every admissible load is of the form:

$$\sigma(x) + \sum_{i=1}^{n} c_i \delta(x - \xi_i) + \sum_{i=1}^{m} d_i \delta'(x - \xi_i) ,$$

where $\sigma(x)$ is a bounded measurable function in Ω.

We improve the normalizing conditions:

(a) $\int_\Omega |\sigma(x)|dx + \sum_{i=1}^{n} |c_i| \leq 1$ (or $\leq C_1$)

(b) $\sum_{i=1}^{m} |d_i| \leq 1.$ (or $\leq C_2$)

There are no other admissible loads.

REFERENCES

[1] J. Bernoulli, Acta Eruditorum, (see in particular pages 268-269), Leipzig, 1696.

[2] G. A. Bliss, Lectures on the calculus of variations, Univ. of Chicago Press, Chicago, Ill., 1946.

[3] E. K. Blum, The calculus of variations, functional analysis and optimal control problems, Topics in optimization, G. Leitman, editor, Vol. 31, Mathematics in Science and Engineering Series, Academic Press, New York, 1967, pp. 415-461.

[4] A. G. Butkovskii and A. Ya. Lerner, "Optimal controls with distributed parameter," Dokl. Akad. Nauk S. S. S. R., No. 134, (1960), pp. 778-781.

[5] ____________, "Optimal control systems with distributed parameters," Avtomatika i Telemekhanika, No. 6, Vol. 21, (1960), pp. 682-691.

[6] A. G. Butkovskii, "Optimum process in systems with distributed parameters," Avtomatika i Telemekhanika, No. 1, Vol. 22, (1961), pp. 17-26.

[7] ____________, "The maximum principle for optimum systems with distributed parameters" Avtomatika i Telemekhanika, No. 10, Vol. 22, (1961), pp. 1288-1301.

[8] ____________ and L. N. Poltavskii, "Optimal control of a distributed oscillatory system", English translation-- Automation and Remote Control, 26, 1965, pp. 1835-1848.

[8a] ____________, Ibid., 27, 1966, pp. 1542-1547.

[8b] ____________, Ibid., 27, 1966, pp. 553-563.

[9] L. Cesari, "Existence theorems for weak and usual optimal solutions in Lagrange problems with unilateral constraints," parts I, II, Trans. Am. Math. Soc. No. 124, (1966), pp. 369-412, 413-430.

[10] A. I. Egorov, "On optimal control of distributed objects," Prikladnaya Matematika i Mechanika, No. 4, Vol. 27, (1963), pp. 683-696.

[11] A. G. Butkovskii, A. I. Egorov, and A. K. Lurie, "Optimal control of distributed systems," SIAM J. on Control, Vol. 6, No. 3, 1968, pp.437-476.

[12] A. I. Egorov, "Optimal processes and invariance theory," SIAM J. on Control, #4 (1966) pp.601-661.

[13] H. O. Fattorini, "Time optimal control of solutions of operational differential equations," Journal of SIAM, Series A on Control, #1, Vol. 2, (1964), pp.54-59.

[14] H. O. Fattorini, "Boundary Control of Systems," SIAM J. on Control, Vol. 6, #3, 1968, pp. 349-385.

[15] I. M. Gel'fand and G. E. Shilov, Generalized Functions, Vol. 1, Academic Press, New York, 1964.

[16] V. Komkov, "A note on the vibration of thin inhomogeneous plates, ZAMM, 48, (1968), pp. 11-16.

[17] ____________, "The optimal control of a transverse vibration of a beam," SIAM J. on Control, #3, Vol. 6, (1968), pp. 401-421.

[18] ____________, "The optimal control of vibrating thin plates," SIAM J. on Control, Vol. 8, #2, (1970), pp. 273-304.

[19] ____________, "A classification of boundary conditions in optimal control theory of elastic systems, to appear.

[20] J. P. LaSalle, "The time optimal control problem," Contributions to non-linear oscillations, Vol. V., Annals of Mathematics, Study #45, Princeton Univ. Press, Princeton, N. J., 1960.

[21] A. E. H. Love, A treatise on the mathematical theory of elasticity, 4th edition Dover Publ., New York, 1944.

[22] E. H. Mansfield, "The bending and stretching of plates" McMillan Co., New York, 1964.

[23] ____________, "On the analysis of elastic plates of variable thickness," Quart. J. Mech. and Appl. Math. 15, (1962), pp. 167-181.

[24] L. S. D. Morley, Skew plates and structures, McMillan Co., New York, 1963.

[25] ___________, "Variational reduction of the clamped plate to two successive membrane problems with an application to a uniformly loaded section," Quart Journal Mech. and Appl. Math., XVI, (1963), pt. 4, pp. 451-471.

[26] N. O. Myklestad, "New method of calculating natural modes of coupled bending-torsional vibrations of beams," Trans. ASME, Vol. 67 (1945), pp. 61-63.

[27] N. I. Muskhelishvili, Some basic problems of the mathematical theory of elasticity, P. Noordhoff, Groningen, Holland, 1953.

[28] L. S. Pontryagin, V. G. Boltyanskii, R. V. Gamkrelidze and E. F. Mishchenko, "The mathematical theory of optimal processes," Interscience Publ. Co., New York, 1962.

[29] ___________, "The theory of optimal processes; the maximum principle," Am. Math. Soc. Translations, Series 2, 18, 1961, pp. 341-382.

[30] ___________, E. F. Mishchenko, A statistical optimal control problem, Izv. Akad. Nauk S.S.S.R., Ser. Mat. 25, (1961), pp.477-498.

[31] E. Reissner, "On bending of elastic plates," Quart. Appl. Math. #5, (1947), pp.395-401.

[32] D. L. Russell, "Linear symmetric hyperbolic systems," SIAM J. on Control, 4, (1966) pp. 276-294.

[33] ___________, "Optimal regulation of linear symmetric hyperbolic systems with finite dimensional controls," M.R.C. Technical Report #566, 1965, Madison, Wisconsin.

[34] D. I. Sherman, "On the solution of plane static problem of the theory of elasticity for given external forces," Dokl. Akad. Nauk S.S.S.R., XXVII, #9, (1940), pp. 907-910.

[35] S. L. Sobolev, Applications of functional analysis in mathematical physics, Am. Math. Soc. Transl., Providence, R. I., 1963.

[36] I. S. Sokolnikoff, Mathematical theory of elasticity, McGraw-Hill, New York, 1956.

[37] S. P. Timoshenko and S. Woinowski-Krieger, The theory of plates and shells, McGraw-Hill, New York, 1960.

[38] __________, History of strength of materials, McGraw-Hill, New York, 1954.

[39] __________ and J. M. Gere, Theory of elastic stability McGraw-Hill, New York, 1961.

[40] C. Wilcox, "Wave operators and assymptotic solutions of wave propagation problems of classical physics," Archive for Rational Mechanics and Analysis, 22, 1966, pp. 37-78.

[41] P. K. C. Wang and F. Tung, "Optimum controls of distributed parameter systems," Proceedings of Joint Control Conference, 1963, pp. 16-31.

[42] K. Yosida, Lectures on differential and integral equations, Interscience, New York, 1960.

[43] L. C. Young, The calculus of variations and optimal control theory, Saunders, New York, 1969.

[44] R. Courant and D. Hilbert, Methods of mathematical physics Interscience, New York, Vol. 1, 1953, Vol. 2, 1962.

[45] J. L. Lions, "Control Optimal de Systemes gouvernes par des Equations aux derivees particles," Gauthier Villars, Paris, 1968.

[46] D. L. Russell,"Boundary Value Control of higher dimensional wave equation" , #1, SIAM J. on Control, Vol 9., 1971.

[46a] __________, __________, #2, Technical Report #95, MRC. University of Wisconsin, 1970.

[47] V. Komkov, The class of admissible loads of the linear plate and beam theory, Instit. Lonib. Acad. Science e lettere, (A) vol. 105, (1971) pp. 329-335.

[48] V. Komkov, Classification of boundary conditions in the optimal control theory of beams, Proc. IFAC Symposium on the control of distributed parameter systems, Banff, Canada, (1971) section 5.5. pp. 1-10.

[49] E. R. Barnes, Necessary and sufficient optimality conditions for a class of distributed parameter control systmes, SIAM J. Control, Vol. 9, #1, (1971), pp. 62-82.

[50] M. Kuehne, Optimal feed back control of flexible mechanical systems, Proc. IFAC Symposium on the control of distirbuted parameter systems, Banff, Canada, (1971), section 12.7.

Lecture Notes in Mathematics

Comprehensive leaflet on request

Vol. 74: A. Fröhlich, Formal Groups. IV, 140 pages. 1968. DM 16,–

Vol. 75: G. Lumer, Algèbres de fonctions et espaces de Hardy. VI, 80 pages. 1968. DM 16,–

Vol. 76: R. G. Swan, Algebraic K-Theory. IV, 262 pages. 1968. DM 18,–

Vol. 77: P.-A. Meyer, Processus de Markov: la frontière de Martin. IV, 123 pages. 1968. DM 16,–

Vol. 78: H. Herrlich, Topologische Reflexionen und Coreflexionen. XVI, 166 Seiten. 1968. DM 16,–

Vol. 79: A. Grothendieck, Catégories Cofibrées Additives et Complexe Cotangent Relatif. IV, 167 pages. 1968. DM 16,–

Vol. 80: Seminar on Triples and Categorical Homology Theory. Edited by B. Eckmann. IV, 398 pages. 1969. DM 20,–

Vol. 81: J.-P. Eckmann et M. Guenin, Méthodes Algébriques en Mécanique Statistique. VI, 131 pages. 1969. DM 16,–

Vol. 82: J. Wloka, Grundräume und verallgemeinerte Funktionen. VIII, 131 Seiten. 1969. DM 16,–

Vol. 83: O. Zariski, An Introduction to the Theory of Algebraic Surfaces. IV, 100 pages. 1969. DM 16,–

Vol. 84: H. Lüneburg, Transitive Erweiterungen endlicher Permutationsgruppen. IV, 119 Seiten. 1969. DM 16,–

Vol. 85: P. Cartier et D. Foata, Problèmes combinatoires de commutation et réarrangements. IV, 88 pages. 1969. DM 16,–

Vol. 86: Category Theory, Homology Theory and their Applications I. Edited by P. Hilton. VI, 216 pages. 1969. DM 16,–

Vol. 87: M. Tierney, Categorical Constructions in Stable Homotopy Theory. IV, 65 pages. 1969. DM 16,–

Vol. 88: Séminaire de Probabilités III. IV, 229 pages. 1969. DM 18,–

Vol. 89: Probability and Information Theory. Edited by M. Behara, K. Krickeberg and J. Wolfowitz. IV, 256 pages. 1969. DM 18,–

Vol. 90: N. P. Bhatia and O. Hajek, Local Semi-Dynamical Systems. II, 157 pages. 1969. DM 16,–

Vol. 91: N. N. Janenko, Die Zwischenschrittmethode zur Lösung mehrdimensionaler Probleme der mathematischen Physik. VIII, 194 Seiten. 1969. DM 16,80

Vol. 92: Category Theory, Homology Theory and their Applications II. Edited by P. Hilton. V, 308 pages. 1969. DM 20,–

Vol. 93: K. R. Parthasarathy, Multipliers on Locally Compact Groups. III, 54 pages. 1969. DM 16,–

Vol. 94: M. Machover and J. Hirschfeld, Lectures on Non-Standard Analysis. VI, 79 pages. 1969. DM 16,–

Vol. 95: A. S. Troelstra, Principles of Intuitionism. II, 111 pages. 1969. DM 16,–

Vol. 96: H.-B. Brinkmann und D. Puppe, Abelsche und exakte Kategorien, Korrespondenzen. V, 141 Seiten. 1969. DM 16,–

Vol. 97: S. O. Chase and M. E. Sweedler, Hopf Algebras and Galois Theory. II, 133 pages. 1969. DM 16,–

Vol. 98: M. Heins, Hardy Classes on Riemann Surfaces. III, 106 pages. 1969. DM 16,–

Vol. 99: Category Theory, Homology Theory and their Applications III. Edited by P. Hilton. IV, 489 pages. 1969. DM 24,–

Vol. 100: M. Artin and B. Mazur, Etale Homotopy. II, 196 Seiten. 1969. DM 16,–

Vol. 101: G. P. Szegö et G. Treccani, Semigruppi di Trasformazioni Multivoche. VI, 177 pages. 1969. DM 16,–

Vol. 102: F. Stummel, Rand- und Eigenwertaufgaben in Sobolewschen Räumen. VIII, 386 Seiten. 1969. DM 20,–

Vol. 103: Lectures in Modern Analysis and Applications I. Edited by C. T. Taam. VII, 162 pages. 1969. DM 16,–

Vol. 104: G. H. Pimbley, Jr., Eigenfunction Branches of Nonlinear Operators and their Bifurcations. II, 128 pages. 1969. DM 16,–

Vol. 105: R. Larsen, The Multiplier Problem. VII, 284 pages. 1969. DM 18,–

Vol. 106: Reports of the Midwest Category Seminar III. Edited by S. Mac Lane. III, 247 pages. 1969. DM 16,–

Vol. 107: A. Peyerimhoff, Lectures on Summability. III, 111 pages. 1969. DM 16,–

Vol. 108: Algebraic K-Theory and its Geometric Applications. Edited by R. M. F. Moss and C. B. Thomas. IV, 86 pages. 1969. DM 16,–

Vol. 109: Conference on the Numerical Solution of Differential Equations. Edited by J. Ll. Morris. VI, 275 pages. 1969. DM 18,–

Vol. 110: The Many Facets of Graph Theory. Edited by G. Chartrand and S. F. Kapoor. VIII, 290 pages. 1969. DM 18,–

Vol. 111: K. H. Mayer, Relationen zwischen charakteristischen Zahlen. III, 99 Seiten. 1969. DM 16,–

Vol. 112: Colloquium on Methods of Optimization. Edited by N. N. Moiseev. IV, 293 pages. 1970. DM 18,–

Vol. 113: R. Wille, Kongruenzklassengeometrien. III, 99 Seiten. 1970. DM 16,–

Vol. 114: H. Jacquet and R. P. Langlands, Automorphic Forms on GL (2). VII, 548 pages. 1970. DM 24,–

Vol. 115: K. H. Roggenkamp and V. Huber-Dyson, Lattices over Orders I. XIX, 290 pages. 1970. DM 18,–

Vol. 116: Séminaire Pierre Lelong (Analyse) Année 1969. IV, 195 pages. 1970. DM 16,–

Vol. 117: Y. Meyer, Nombres de Pisot, Nombres de Salem et Analyse Harmonique. 63 pages. 1970. DM 16,–

Vol. 118: Proceedings of the 15th Scandinavian Congress, Oslo 1968. Edited by K. E. Aubert and W. Ljunggren. IV, 162 pages. 1970. DM 16,–

Vol. 119: M. Raynaud, Faisceaux amples sur les schémas en groupes et les espaces homogènes. III, 219 pages. 1970. DM 16,–

Vol. 120: D. Siefkes, Büchi's Monadic Second Order Successor Arithmetic. XII, 130 Seiten. 1970. DM 16,–

Vol. 121: H. S. Bear, Lectures on Gleason Parts. III, 47 pages. 1970. DM 16,–

Vol. 122: H. Zieschang, E. Vogt und H.-D. Coldewey, Flächen und ebene diskontinuierliche Gruppen. VIII, 203 Seiten. 1970. DM 16,–

Vol. 123: A. V. Jategaonkar, Left Principal Ideal Rings. VI, 145 pages. 1970. DM 16,–

Vol. 124: Séminare de Probabilités IV. Edited by P. A. Meyer. IV, 282 pages. 1970. DM 20,–

Vol. 125: Symposium on Automatic Demonstration. V, 310 pages. 1970. DM 20,–

Vol. 126: P. Schapira, Théorie des Hyperfonctions. XI, 157 pages. 1970. DM 16,–

Vol. 127: I. Stewart, Lie Algebras. IV, 97 pages. 1970. DM 16,–

Vol. 128: M. Takesaki, Tomita's Theory of Modular Hilbert Algebras and its Applications. II, 123 pages. 1970. DM 16,–

Vol. 129: K. H. Hofmann, The Duality of Compact Semigroups and C*-Bigebras. XII, 142 pages. 1970. DM 16,–

Vol. 130: F. Lorenz, Quadratische Formen über Körpern. II, 77 Seiten. 1970. DM 16,–

Vol. 131: A. Borel et al., Seminar on Algebraic Groups and Related Finite Groups. VII, 321 pages. 1970. DM 22,–

Vol. 132: Symposium on Optimization. III, 348 pages. 1970. DM 22,–

Vol. 133: F. Topsøe, Topology and Measure. XIV, 79 pages. 1970. DM 16,–

Vol. 134: L. Smith, Lectures on the Eilenberg-Moore Spectral Sequence. VII, 142 pages. 1970. DM 16,–

Vol. 135: W. Stoll, Value Distribution of Holomorphic Maps into Compact Complex Manifolds. II, 267 pages. 1970. DM 18,–

Vol. 136: M. Karoubi et al., Séminaire Heidelberg-Saarbrücken-Strasbourg sur la K-Théorie. IV, 264 pages. 1970. DM 18,–

Vol. 137: Reports of the Midwest Category Seminar IV. Edited by S. MacLane. III, 139 pages. 1970. DM 16,–

Vol. 138: D. Foata et M. Schützenberger, Théorie Géométrique des Polynômes Eulériens. V, 94 pages. 1970. DM 16,–

Vol. 139: A. Badrikian, Séminaire sur les Fonctions Aléatoires Linéaires et les Mesures Cylindriques. VII, 221 pages. 1970. DM 18,–

Vol. 140: Lectures in Modern Analysis and Applications II. Edited by C. T. Taam. VI, 119 pages. 1970. DM 16,–

Vol. 141: G. Jameson, Ordered Linear Spaces. XV, 194 pages. 1970. DM 16,–

Vol. 142: K. W. Roggenkamp, Lattices over Orders II. V, 388 pages. 1970. DM 22,–

Vol. 143: K. W. Gruenberg, Cohomological Topics in Group Theory. XIV, 275 pages. 1970. DM 20,–

Vol. 144: Seminar on Differential Equations and Dynamical Systems, II. Edited by J. A. Yorke. VIII, 268 pages. 1970. DM 20,–

Vol. 145: E. J. Dubuc, Kan Extensions in Enriched Category Theory. XVI, 173 pages. 1970. DM 16,–

Please turn over

Vol. 146: A. B. Altman and S. Kleiman, Introduction to Grothendieck Duality Theory. II, 192 pages. 1970. DM 18,

Vol. 147: D. E. Dobbs, Cech Cohomological Dimensions for Commutative Rings. VI, 176 pages. 1970. DM 16,

Vol. 148: R. Azencott, Espaces de Poisson des Groupes Localement Compacts. IX, 141 pages. 1970. DM 16,-

Vol. 149: R. G. Swan and E. G. Evans, K-Theory of Finite Groups and Orders. IV, 237 pages. 1970. DM 20,-

Vol. 150: Heyer, Dualität lokalkompakter Gruppen. XIII, 372 Seiten. 1970. DM 20,-

Vol. 151: M. Demazure et A. Grothendieck, Schémas en Groupes I. (SGA 3). XV, 562 pages. 1970. DM 24,-

Vol. 152: M. Demazure et A. Grothendieck, Schémas en Groupes II. (SGA 3). IX, 654 pages. 1970. DM 24,-

Vol. 153: M. Demazure et A. Grothendieck, Schémas en Groupes III. (SGA 3). VIII, 529 pages. 1970. DM 24,-

Vol. 154: A. Lascoux et M. Berger, Variétés Kähleriennes Compactes. VII, 83 pages. 1970. DM 16,-

Vol. 155: Several Complex Variables I, Maryland 1970. Edited by J. Horváth. IV, 214 pages. 1970. DM 18,-

Vol. 156: R. Hartshorne, Ample Subvarieties of Algebraic Varieties. XIV, 256 pages. 1970. DM 20,-

Vol. 157: T. tom Dieck, K. H. Kamps und D. Puppe, Homotopietheorie. VI, 265 Seiten. 1970. DM 20,-

Vol. 158: T. G. Ostrom, Finite Translation Planes. IV. 112 pages. 1970. DM 16,-

Vol. 159: R. Ansorge und R. Hass. Konvergenz von Differenzenverfahren für lineare und nichtlineare Anfangswertaufgaben. VIII, 145 Seiten. 1970. DM 16,-

Vol. 160: L. Sucheston, Constributions to Ergodic Theory and Probability. VII, 277 pages. 1970. DM 20,-

Vol. 161: J. Stasheff, H-Spaces from a Homotopy Point of View. VI, 95 pages. 1970. DM 16,-

Vol. 162: Harish-Chandra and van Dijk, Harmonic Analysis on Reductive p-adic Groups. IV, 125 pages. 1970. DM 16,-

Vol. 163: P. Deligne, Equations Différentielles à Points Singuliers Reguliers. III, 133 pages. 1970. DM 16,-

Vol. 164: J. P. Ferrier, Seminaire sur les Algebres Complètes. II, 69 pages. 1970. DM 16,-

Vol. 165: J. M. Cohen, Stable Homotopy. V, 194 pages. 1970. DM 16,-

Vol. 166: A. J. Silberger, PGL_2 over the p-adics: its Representations, Spherical Functions, and Fourier Analysis. VII, 202 pages. 1970. DM 18,-

Vol. 167: Lavrentiev, Romanov and Vasiliev, Multidimensional Inverse Problems for Differential Equations. V, 59 pages. 1970. DM 16,-

Vol. 168: F. P. Peterson, The Steenrod Algebra and its Applications: A conference to Celebrate N. E. Steenrod's Sixtieth Birthday. VII, 317 pages. 1970. DM 22,-

Vol. 169: M. Raynaud, Anneaux Locaux Henséliens. V, 129 pages. 1970. DM 16,-

Vol. 170: Lectures in Modern Analysis and Applications III. Edited by C. T. Taam. VI, 213 pages. 1970. DM 18,-

Vol. 171: Set-Valued Mappings, Selections and Topological Properties of 2^X. Edited by W. M. Fleischman. X, 110 pages. 1970. DM 16,-

Vol. 172: Y.-T. Siu and G. Trautmann, Gap-Sheaves and Extension of Coherent Analytic Subsheaves. V, 172 pages. 1971. DM 16,-

Vol. 173: J. N. Mordeson and B. Vinograde, Structure of Arbitrary Purely Inseparable Extension Fields. IV, 138 pages. 1970. DM 16,-

Vol. 174: B. Iversen, Linear Determinants with Applications to the Picard Scheme of a Family of Algebraic Curves. VI, 69 pages. 1970. DM 16,-

Vol. 175: M. Brelot, On Topologies and Boundaries in Potential Theory. VI, 176 pages. 1971. DM 18,-

Vol. 176: H. Popp, Fundamentalgruppen algebraischer Mannigfaltigkeiten. IV, 154 Seiten. 1970. DM 16,-

Vol. 177: J. Lambek, Torsion Theories, Additive Semantics and Rings of Quotients. VI, 94 pages. 1971. DM 16,-

Vol. 178: Th. Bröcker und T. tom Dieck, Kobordismentheorie. XVI, 191 Seiten. 1970. DM 18,-

Vol. 179: Seminaire Bourbaki – vol. 1968/69. Exposés 347-363. IV, 295 pages. 1971. DM 22,-

Vol. 180: Séminaire Bourbaki – vol. 1969/70. Exposés 364-381. IV, 310 pages. 1971. DM 22,-

Vol. 181: F. DeMeyer and E. Ingraham, Separable Algebras over Commutative Rings. V, 157 pages. 1971. DM 16,-

Vol. 182: L. D. Baumert. Cyclic Difference Sets. VI, 166 pages. 1971. DM 16,-

Vol. 183: Analytic Theory of Differential Equations. Edited by P. F. Hsieh and A. W. J. Stoddart. VI, 225 pages. 1971. DM 20,-

Vol. 184: Symposium on Several Complex Variables, Park City, Utah, 1970. Edited by R. M. Brooks. V, 234 pages. 1971. DM 20,-

Vol. 185: Several Complex Variables II, Maryland 1970. Edited by J. Horváth. III, 287 pages. 1971. DM 24,-

Vol. 186: Recent Trends in Graph Theory. Edited by M. Capobianco/ J. B. Frechen/M. Krolik. VI, 219 pages. 1971. DM 18,-

Vol. 187: H. S. Shapiro, Topics in Approximation Theory. VIII, 275 pages. 1971. DM 22,-

Vol. 188: Symposium on Semantics of Algorithmic Languages. Edited by E. Engeler. VI, 372 pages. 1971. DM 26,-

Vol. 189: A. Weil, Dirichlet Series and Automorphic Forms. V, 164 pages. 1971. DM 16,

Vol. 190: Martingales. A Report on a Meeting at Oberwolfach, May 17-23, 1970. Edited by H. Dinges. V, 75 pages. 1971. DM 16,-

Vol. 191: Séminaire de Probabilités V. Edited by P. A. Meyer. IV, 372 pages. 1971. DM 26,-

Vol. 192: Proceedings of Liverpool Singularities – Symposium I. Edited by C. T. C. Wall. V, 319 pages. 1971. DM 24,-

Vol. 193: Symposium on the Theory of Numerical Analysis. Edited by J. Ll. Morris. VI, 152 pages. 1971. DM 16,-

Vol. 194: M. Berger, P. Gauduchon et E. Mazet. Le Spectre d'une Variété Riemannienne. VII, 251 pages. 1971. DM 22,-

Vol. 195: Reports of the Midwest Category Seminar V. Edited by J. W. Gray and S. Mac Lane. III, 255 pages. 1971. DM 22,-

Vol. 196: H-spaces – Neuchâtel (Suisse)- Août 1970. Edited by F. Sigrist. V, 156 pages. 1971. DM 16,-

Vol. 197: Manifolds – Amsterdam 1970. Edited by N. H. Kuiper. V, 231 pages. 1971. DM 20,-

Vol. 198: M. Hervé, Analytic and Plurisubharmonic Functions in Finite and Infinite Dimensional Spaces. VI, 90 pages. 1971. DM 16,-

Vol. 199: Ch. J. Mozzochi, On the Pointwise Convergence of Fourier Series. VII, 87 pages. 1971. DM 16,-

Vol. 200: U. Neri, Singular Integrals. VII, 272 pages. 1971. DM 22,-

Vol. 201: J. H. van Lint, Coding Theory. VII, 136 pages. 1971. DM 16,-

Vol. 202: J. Benedetto, Harmonic Analysis on Totally Disconnected Sets. VIII, 261 pages. 1971. DM 22,-

Vol. 203: D. Knutson, Algebraic Spaces. VI, 261 pages. 1971. DM 22,-

Vol. 204: A. Zygmund, Intégrales Singulières. IV, 53 pages. 1971. DM 16,-

Vol. 205: Séminaire Pierre Lelong (Analyse) Année 1970. VI, 243 pages. 1971. DM 20,-

Vol. 206: Symposium on Differential Equations and Dynamical Systems. Edited by D. Chillingworth. XI, 173 pages. 1971. DM 16,-

Vol. 207: L. Bernstein, The Jacobi-Perron Algorithm – Its Theory and Application. IV, 161 pages. 1971. DM 16,-

Vol. 208: A. Grothendieck and J. P. Murre, The Tame Fundamental Group of a Formal Neighbourhood of a Divisor with Normal Crossings on a Scheme. VIII, 133 pages. 1971. DM 16,-

Vol. 209: Proceedings of Liverpool Singularities Symposium II. Edited by C. T. C. Wall. V, 280 pages. 1971. DM 22,-

Vol. 210: M. Eichler, Projective Varieties and Modular Forms. III, 118 pages. 1971. DM 16,-

Vol. 211: Théorie des Matroïdes. Edité par C. P. Bruter. III, 108 pages. 1971. DM 16,-

Vol. 212: B. Scarpellini, Proof Theory and Intuitionistic Systems. VII, 291 pages. 1971. DM 24,-

Vol. 213: H. Hogbe-Nlend, Théorie des Bornologies et Applications. V, 168 pages. 1971. DM 18,-

Vol. 214: M. Smorodinsky, Ergodic Theory, Entropy. V, 64 pages. 1971. DM 16,-